Economics of Banana Production and Marketing in the Tropics

(A Case Study of Cameroon)

Esendugue Gregory Fonsah
Angus S.N.D Chidebelu

Langaa Research & Publishing CIG
Mankon, Bamenda

Publisher
Langaa RPCIG
Langaa Research & Publishing Common Initiative Group
P.O. Box 902 Mankon
Bamenda
North West Region
Cameroon
Langaagrp@gmail.com
www.langaa-rpcig.net

Distributed in and outside N. America by African Books Collective
orders@africanbookscollective.com
www.africanbookcollective.com

ISBN: 9956-726-54-0

DISCLAIMER
All views expressed in this publication are those of the author and do not necessarily reflect the views of Langaa RPCIG.

About the Authors

Chief/Dr. Esendugue Greg Fonsah is an Associate Professor and Extension Economist in the Department of Ag & Applied Economics, College of Agriculture and Environmental Sciences, University of Georgia, Tifton Campus, Georgia, USA and an international consultant. His areas of interest are in Farm management, Production Economics, Agricultural Marketing and Consumer Demand, Agribusiness Management, International Trade and Policy of fruits and vegetables. Professor Fonsah has taught economics, business mathematics, international business and trade, economic geography and research methodology courses at various institutions of higher learning around the world including the University of Buea, Cameroon, Central China Agricultural University, P.R. of China, Wuhan Institute of Technology, P.R. of China, Kentucky College of Business and the University of Georgia in the United States respectively. Professor Fonsah has written and published over 300 scholarly works; journal articles, bulletins, books, book chapters etc. and participated in close to $10 million research grant as Principal or Co-Principal Investigator.

Despite his passion for academia, the greater part of his professional career was in the corporate world working with Del Monte Fresh Produce, Cameroon, Lapanday Food Co., Philippines, and Aloha Farm Inc, Hawaii, USA. Dr. Fonsah's experience in international agriculture and international food industry has developed extensively in the past 23 years where he served in various senior managerial positions in Fresh Food Multinational Corporations in Africa, Asia and North America respectively. He obtained his B.S. in Business Management from Berea College, his MBA (Masters in Business Administration) from Morehead State University, MS in Agricultural Economics from the University of Kentucky and Ph.D. in Agricultural Economics from the University of Nigeria, Nsukka, Nigeria, West Africa respectively. He and his wife, Liu Kailan have two children, Derrick Bebongho and Leilani Ebongkie-Jean.

Professor Sonny Angus N. D. Chidebelu was educated at the University of Nigeria, Nsukka, Nigeria; the University of Guelph, Guelph, Ontario, Canada; and the University of Georgia, Athens, Georgia, U.S.A. At the University of Nigeria, he was: Head of the Department of Agricultural Economics; Associate Dean, Faculty of Agriculture; and Associate Dean, School of Postgraduate Studies. He has consulted for National and

International Organizations and has over 50 publications. He is married with four children.

Table of Contents

List of Tables

List of Figures

Preface

In most African countries, banana production has been consigned to subsistence production with little attempt at developing it. However, a few countries, especially the Francophones in West Africa, have recognised its commercial importance and used their special relationship with France to export bananas. This has led to the dualization of the banana sector, with the traditional system existing side by side with a modern sector geared towards export trade.

This book is one of the few comprehensive studies that has incorporated both the agronomic and economic aspects of banana production and marketing in Africa. It has looked at all facets of banana production, from an historical perspective to the various adopted technologies, namely: the traditional, the semi-modern, and the modern production system respectively. Conversely, the marketing aspect covers both the domestic and world markets (international trade), with emphasis on the preferential (ACP / DOM Lome Convention) and the open markets of the European Economic Community (EEC).

Unfortunately most texts in banana production have ignored the internationalisation of the banana trade and the ever-increasing investment portfolio geared towards this crop, which is not only the backbone of several Third World economies, but has subsequently raised the awareness of a large body of serious tropical students to carry out research.

As a result we have endeavoured to update this book both descriptively and analytically by adding three chapters on research methods and two chapters on profitability analysis. With these provisions we believe this book will be useful to agronomists, entomologists, economists, farm managers, financial analysts, business investors, marketing specialists, government policy makers, large, medium and small scale banana growers, college and/or university teachers.

Although the emphasis is placed on Cameroon, other relevant African, tropical and subtropical banana-producing countries are mentioned where necessary, especially in the export sector where a degree of competition existed. Further, agricultural practices, soils, meteorological and climatological characteristics, pests and diseases, personnel and banana varieties grown, mean that findings in Cameroon are of relevance to other banana-producing countries, especially in Africa. Meanwhile, other African and tropical countries still contemplating entry into banana exports are advised to read this book to benefit from the Cameroon experience.

We hope this book will be read by university students especially those specialising in agriculture, trade, marketing, economic history and agro-business.

Esendugue Gregory Fonsah, Ph.D
Angus S. N. D. Chidebelu, Ph.D

Acknowledgement

Special acknowledgements are due to many people who assisted either in the earlier or later stages of this book. At the University of Nigeria, Nsukka, special thanks go to Dr. Ethelbert C Nwagbo, who was one of the supervisors of a doctoral dissertation, on which this work is based. We are equally indebted to Dr. Manfred T. Bessong, Deputy Director of Ekona Research Centre (IRA), Cameroon, for his contributions throughout the initial and later stages of this work. Many thanks to several Faculty members and graduate students who assisted in one way or the other during the initial development of this book.

We wish to express our gratitude to the Del Monte staff who contributed in various ways especially Dr. Charles Agho Adamu, for reviewing Chapters Four and Five. Thanks to Jack Dwyer, the former General Manager of Del Monte Cameroon, Luis Valerio, Luis Sanchez, Luis Martinez, Ray Galindo, the late Ernesto Piniero, Liz Parr, Ann Kanyi and Victor Range. Equal words of thanks to Mathias Bena, Project Manager, CDC/ Del Monte Tiko Banana Project, Oscar Ndoh, Estate Manager, Ekona Banana Plantation, and Aboubakar Perfoura who assisted in the development of Part Five. Thanks to Fonge Ambroise for editing the page proofs, Dr. Olpoc and Dennis Tambe for various contributions.

Finally, our eternal love, deepest gratitude and affection go to our beloved wives, Kailan Liu Fonsah and Felicia Ijeoma Chidebelu for their continuous encouragement, moral support, patience and understanding exercised during these extremely difficult years.

Esendugue Gregory Fonsah, Ph.D
Angus S. N. D. Chidebelu, Ph.D

Part One

Cameroon's Economy And The Agricultural Sector

Chapter One

Introduction

For the past half a century Cameroon's banana industry has continually adopted either the semi-modern or traditional production technologies which, consequently, has either resulted in low yields, increased costs or low profits. The industry recorded only three percent out of 33% of total agricultural commodities exported in 1980/1981. This figure is relatively insignificant compared to the total of approximately 3,252 hectares (ha) of banana cultivated land. Further, records in 1984 depicted a total of 57,900 mt of banana shipped to foreign countries representing an average yield of 17.8 mt/ha (FAO, 1986).

Empirical studies in Cameroon, Central America and the Philippines have shown that a minimum average yield of 40 mt/ha could be expected during the first year of production in a modern plantation. This quantity could be increased from the second and subsequent years (Stover and Simmond, 1987; FAO, 1986). The adoption of modern banana technology in Cameroon would increase yield drastically, but it is risky and very costly (Knutson et al, 1983; Eicher and Staatz, 1984; Gittinger, 1984).

Further, assuming yields increase, will there be other marketing outlets for Cameroon since the currently existing market can only absorb a maximum of 60,000 rat/annum, ceteris paribus (Lome Convention, 28th Feb., 1975)? One logical answer will be to dispose of the excess products through other marketing channels. The non-banana producing African and European countries could be targeted, or strive for better quality, and become competitive in the open market. The question - should the Cameroon banana industry adopt modern technology or stay with its traditional and semi-modern methods? - is yet to be solved.

The Cameroon banana marketing trend for the past decade is downward sloping. Decreasing yields, due to increase disease infestation and lack of appropriate irrigation system, phytosanitary treatment, and cableway network have all partially contributed to the deteriorating quality and exportable output (FAO, 1986).

Thus, the problem being tackled is essentially the profitability of the adoption of improved banana production technologies in Cameroon and the concomitant issue of the disposal of increased output.

1.1 Background on Cameroon

Cameroon obtained her independence from France and Britain on 1st January, 1960. Since then the country has been one of the most politically stable in the continent of Africa. Besides its political stability the country operates on "a unified political party system known as The Cameroon People's Democratic Movement (CPDM) and the legal system is based on French and English laws" (Nelson et al, 1974).

Cameroon is located in central Africa, according to the UN zoning, and occupies an area of 475,000 km². From north to south it is 1,120 km whereas the east-west distance is 720 km (Nelson et al, 1974). The country's close neighbours are Nigeria to the west, Central African Republic and Chad to the east, Lake Chad to the north, and Congo, Gabon, and Equatorial Guinea to the south (Appendix HI). The 1991 projected population of 12,243,700 (Neba, 1987), is made up of 200 ethnic groups (tribes) living in ten provinces, namely: Far North, North, Adamawa, East, Centre, South, Littoral, Southwest, Northwest, and West (Appendix III). The people of Cameroon practise Christianity, mostly in the south, and traditional religions and Islam, mostly in the north. The climate varies from equatorial in the south to arid in the north with sporadic cool mountain climates. The current literacy rate is 54% (World Bank Atlas 1992; Britannica, 1992), compared to 20% in 1977 (The World Almanac, 1982).

1.2 Economic Growth

According to the USDA's Report of July 1985 Cameroon was plagued by a depressing economic situation in 1982. However, this situation was overturned in 1984 when the economy picked up again due to "normal rainfall". The quantity of food and cash crops produced hiked dramatically. Further, "world prices for most of Cameroon's agricultural exports rose, and oil export earnings increased as a larger export volume offset a decline in world prices. Real GDP increased 6.5 percent in 1983/84, compared with 4.8 percent in 1982/83".

4

The country has enjoyed a positive economic growth since her independence in 1960 for two significant reasons, namely: the country is rich in natural resources, and she is politically stable. During the initial decade of independence the Gross Domestic Product (GDP) rose by 2.8 percent. This figure has almost tripled in the '80s as a result of oil sales. Cameroon is classified as a "middle-income country" as the per capita income rose to $829 in the 80s - hence, a 0.5 percent increase in "real personal consumption expenditure" (USDA, July 1985).

1.3 The Agricultural Sector

Cameroon's agricultural sector includes forestry, animal husbandry and fishing. As Burfisher (1984) noted, 83% of the working population was absorbed by this sector in 1984. Moreover, agriculture contributed 33% of the total revenue amassed from goods sold to other countries in the fiscal year 1980/81. It also contributed 28% of GDP.

Cameroon's major food crops are: cassava, yams, cocoyams (taro-macabo), plantains, maize, millet, and sorghum. For the past decade Cameroon has been among the top five exporters of cocoa. It has also a made significant contribution in the sale of coffee from the African continent. Besides cotton and banana, Cameroon also exports wood and rubber (Burfisher, 1984).

Out of Cameroon's 12.2 million inhabitants (Neba, 1987) about 70% live in the rural areas and approximately 90% of the rural dwellers are traditional farmers (USDA, July 1985).

According to the USDA Report (July 1985) the range of farm income was between $160-$260 in 1982. This amount is relatively low compared to the non-farm sector and/or the national average of $829. This substantial gap between farm and non-farm income has attracted some farmers into the non-farm sectors (Knutson, et al, 1983). Also, since the construction of oil refineries, the significance of agriculture has diminished as its share in the 1983/84 GDP dropped to 20% compared to 38% in 1979/80. Moreover, revenue from agricultural products sold to other countries dropped to 20% in 1983/84 compared to 41% in 1979/80.

Burfisher (1984) stated that Cameroon's agriculture has been successful for the following reasons: (I) its favourable climate which enables the cultivation of diverse food and cash crops; (ii) the concentration of the gifts

5

of nature, irrespective of its size, and the fact that the gift of nature has been haphazardly distributed all over the world, yet Cameroon is rich in natural resources, and (iii) most importantly, is the "government's policy of supporting the agricultural sector".

The government determines prices for certain food and cash crops. This includes rice, cocoa, coffee, cotton, palm kernels, groundnuts, tobacco, palm oil and rubber. The prices are regulated on the tenth month of each year. In the past decade the regulated producer price has always been less than the world price (Burfisher, 1984).

Studies have proven that the producer price policy was a tool that the Cameroon government adopted to raise revenue from farmers, which she injected into other programmes. "Surpluses generated by the differences between producer prices and world prices provided funds to Cameroon's monopoly crop marketing agency (NPMB). The funds, supposedly, were used to support a price stabilisation policy, to finance rural infrastructural developments, and to make contributions to the national budget for the development of non-agricultural sectors" (Burfisher, 1984; Nelson et al, 1974).

It is unquestionable that the producer price policy fetched a considerable amount of revenue to the government of Cameroon during 1976-78 when world prices skyrocketed for cash crops, especially because the excess profit from the transaction did not reach the farmers. After analysing the situation, one is tempted to partially accredit the economic growth in the non-farm region to the farmers during this period.

Since 1979, the government has continually raised the ceiling price for food and cash crops. As Burfisher (1984) puts it, "Producer price increases have equalled or exceeded inflation". In order to encourage the depressed farmers a new policy has been enacted. It is believed that if "real farm income" is elevated, the chances are that there would be a positive response in the cultivation of agricultural food commodities. So far, only farmers involved in the cultivation of rice are enjoying a price beyond the world price. This policy is designed to reduce the quantity of rice purchased from outsiders by Cameroonians as their purchasing power for rice has increased.

Chapter Two

Historical Perspective

2.1 Banana production in the former West Cameroon

In the former West Cameroon, the first banana plantation was created in 1907 by Afrikanische Frucht A. G. of Hamburg (AFC). It was 5,000 ha large and located around the Tiko plain, Fako, the former West Cameroon. Then, only the local variety was planted and exported to Germany as dehydrated banana. All shipment was done by *Woermann Lime,* a German vessel, through the Tiko wharf, also built by the AFC in 1913 at Keka Island. It was only in 1910 that two new varieties, 'Gros Michel' and 'Sinensis' (Grande Maine and Poyo), from Costa Rica and the Canary Islands, were introduced (Lecoq, 1972; Epale, 1985).

After the First World War the German growers repurchased their plantation in 1930 because the British businessmen were not interested in buying them (Ngoh, 1987). Table 2.1 depicts that 10,000 mts were exported in 1931. During this same time the first Cameroon banana plantation (Compagnie des Bananes) under French auspices was established in the Mungo Division of the former East Cameroon (Lecoq, 1972).

Table 2.1: Banana Export: Cameroon to Britain - 1931-1938

Year	Quantity (mts)
1931	10,000
1932	15,540
1933	17,600
1934	23,690
1935	39,110
1936	50,860
1937	57,270
1938	57,640

Source: Lecoq (1972), L'Evolution de l'Economie Baniere au Cameroun. Fruits, Vol. 27, No. 10, pp. 677-696.

Between 1884-1914 several German plantations were established in the former West Cameroon (Appendix I), cultivating principally palm products, cocoa, rubber, bananas, kola nuts, cotton, coffee, tobacco and sylvan products (Table 2.2). Furthermore, Table 2.2 shows that approximately two million mature and immature banana plants were sown on a total area of 2,559 ha in 1931.

Table 2.2: Extent of Cultivation in the Plantations, 1913

Crops	Area cultivated (hectares)	Mature area (hectares)	No. of plants
Oil Palm	5,044	1,647	1,257,569
Banana	2,164	395	1,921,345
Cocoa	13,161	8,175	7,791,345
Coffee	107	10	115,564
Tobacco	153	92	2,052,000
Ficus	43	8	41,150
Kickxia	3,588	966	4,696,909
Cassava	175	29	116,721
"Lianes"	-	-	20,000
Hevea	3,589	-	1,143,803
"Castilia"	7	1	2,584
Sisal	30	22	26,015
Total	28,061	11,345	19,184,738

Source: Deutsche Schutagebiete (1913), quoted in Labouret; Le Cameroon (Paris, 1037), p. 164 and Epale, J. S. (1985), Plantations and Development in Western Cameroon, 1885-1975, Vintage Press Inc., New York, p. 44.

After the declaration of World War One (WWI) in 1914, the Germans lost to the joint French and English troops. Their plantations were seized under the "Custodian of Enemy Property" pretext. The former West Cameroon was given to the British, whilst the former East Cameroon was given to the French. This apportionment and transition led to the Treaty of Versailles in 1919 (Lecoq, 1972; Epale, 1985).

After World War II the German plantation was again sequestered and donated to the Cameroon Development Corporation (CDC). CDC in turn leased it out to a British company, Elders and Fyffes, affiliated to the United Fruit Company (UFC), with its head office in Boston, U.S.A. Compagnie des Bananes in the former East Cameroon was also an affiliate of The United Fruit Company. In 1947 15,376 mts were exported to Britain, out of which 11,000 rats belonged to CDC (Lecoq, 1972; Ngoh, 1987).

Elders and Fyffes was the only banana supplier from Africa in the British market. Per capita consumption in Britain was 3 kg in 1947 compared to 9 kg in 1938. This situation changed at the end of 1947 as a Jamaican-based British banana plantation was completely destroyed by the Panama Disease (fusarial wilt). Consequently, Britain requested an additional 20,000 mts from the French producers. Belgium, Switzerland, and other Scandinavian countries did the same for fear that Elders and Fyffes might not meet their demand. In January 1948, the CFA franc was pegged to the French franc at 50 FCFA = 1 FF (Nelson et al, 1974; Cunha, 1984). This change in the Cameroon monetary system encouraged the banana exports to Europe as it facilitated currency convertibility.

According to Lecoq (1972) various banana organisations were established between 1947 and 1972. Elders and Fyffes (UFC) represented CDC as producer, buyer, and transporter. On 5th August, 1952 the first banana cooperative was formed, the Bakwerie Cooperative Union of Farmers Ltd (BCUF). With government assistance, this cooperative exported its own bananas directly, thus refusing to accept CDC as intermediary. In 1960, another cooperative was formed in the Kumba region, the Kumba Federation of Cooperative and Produce Marketing Societies (K/F of CPMS Ltd), and in May 1967, a new one was created in Muyuka, the Progressive Farmers' Produce Enterprise (PFPC).

A breakdown of quantities exported by the smallholders and cooperative producers from 1960-1964 is illustrated in Table 2.3. It shows that 74% and 69% of total exports from non-CDC sources were produced by BCUF in 1960 and 1964 as compared to 13% and 2% by K/F in 1960 and 1964, respectively. Unfortunately, production by the cooperatives and smallholders dropped to zero percent in 1971. The transition from bunches to boxed bananas began in 1961 when the first packing station was constructed in Molyko. However, the initial trial was done in the Fyffes Plantation at Bavenga, Fako Division. By total production had been completely converted to boxes. In 1964, West Cameroon gained an Italian market to supply bananas bi-monthly at the cost, insurance, freight (GIF) price of £65/mt. In the former East Cameroon producer allocated part of their French market quota of 12,000 mts to the former West Cameroon producers since they could not meet the target due to declining production.

Table 2.3: Banana Export (mts) by Various non-CDC Producers

Year	BCUF	K/F	Smallholder	Fyffes	Total
1960	37,500	6,450	4,700	2,300	50,950
1961	27,070	7,830	9,660	5,360	49,920
1962	27,680	6,020	8,970	5,110	47,780
1963	21,120	3,950	4,390	6,375	35,835
1964	20,200	630	3,950	4,570	29,350

Source: Lecoq (1972), L'Evolution de l'Economie Bananiere au Cameroun. Fruits, Vol. 27, No. 10, pp. 677-686.

2.2 Banana production in the former East Cameroon

Under the French mandate, the former East Cameroon began banana production in 1931 at Penja with two farms, owned by Messrs Adolphe Beynis and Jacqueys, and at Mbanga, by Messrs Maurice Beynis and Rossides (Appendix II). Both plantations started off successfully with the "Gros Michel" variety. *Woermann Linie,* the German shipping line that transported the Afrikanische Frucht from Tiko to Europe, was also contracted until 1935, when two French shipping companies, L. Martin, and The Maritime Transport Company, CTM, created branches in Cameroon. By the end of 1936, all export was done through the French shipping companies, La Companie des Chargeurs Reunis and CTM. By 1939, export volume had increased to 26,753 mts (Table 2.4), out of which 96% penetrated the French market and only 4 percent was channelled to other European countries.

Table 2.4: Bananas Exported to Europe, 1932-1939

Years	To France	Other Countries in Europe	Total Expenditure
1932	-	-	-
1933	93	500	593
1934	1521	350	1871
1935	7,648	82	7,730
1936	16,310	555	16,865
1937	21,340	2,794	24,134
1938	25,236	756	25,992
1939	25,721	1,032	26,753

Source: Lecoq (1972), L'Evolution de l'Economie Bananiere au Cameroun. Fruits, Vol. 27, No. 10, pp. 677-696.

Although banana production in the former East Cameroon began in 1931, it was not until 1936 that a union was created, "Syndicat de Defense des Interet Bananiers du Cameroun" (SDIBC). Only European growers were represented in the banana union since Cameroonians were not interested in export banana production. In 1946, small-scale production was encouraged, and in 1950 cooperatives and other agricultural firms, "Societes Agricoles de Prevoyance" (SAP) came into existence. Table 2.5 shows the breakdown of industrial and small-scale production from 1946 to 1959.

It further depicts that from 1952 to 1959, more than 50% of total output from the former East Cameroon was produced by small-scale farmers.

In 1964, the Cooperative Union, "Union Generale des Cooperatives Bananiers du Cameroun" (UGECOBAM), representing both the industrial and small-scale European and Cameroonian affiliates (Table 2.6), was formed. The industrial firms were all members of "SOBACO" North, Central, and South.

Table 2.5 Industrial and Small-scale Export Volume (mts), 1946-1959

Year	Industrial	Small-Scale	Total
1946	6,900	300	7,200
1947	17,200	2,900	20,100
1948	25,600	5,700	31,300
1949	26,300	8,800	35,100
1950	32,500	16,600	49,100
1951	36,600	19,100	55,700
1952	22,900	26,900	49,800
1953	38,600	38,300	76,900
1954	36,800	37,300	74,100
1955	31,800	41,600	73,400
1956	23,100	43,000	66,100
1957	31,500	53,700	85,200
1958	21,700	51,700	73,400
1959	27,700	32,500	60,200

Source: Lecoq (1972), L'Evolution de l'Economie Bananiere au Cameroun. Fruits, Vol. 27, No. 10, pp. 677-696.

Table 2.6 Cooperatives affiliated to UGECOBAM, 1964

Names	No. of Workers		Total Output
	Europeans	Cameroonians	(mts)
SOBACO North	6	2	7,130
SOBACO South	8	2	6,160
SOBACO Central	10		9,800
UNICOPRODICAM		1,000	8,900
UCAPC		400	5,700
COFRUICAM		500	5,350
UBM		500	3,380
COOPLABAM		350	8,900
SAP (not affiliated)		400	8,500

Source: Lecoq (1972), L'Evolution de l'Economie Bananiere au Cameroun. Fruits, Vol. 27, No. 10, pp. 677-696.

By 1971, only the three SOBACO cooperatives and a few individual producers (Table 2.7) were still in the export business. The other cooperative unions (Table 2.6) had exited for one or a combination of the following reasons:

- Panama Disease infestation of the Mbanga region in 1954,
- Sigatoka infestation in 1956,
- Political upheaval in 1959,
- Change from the local to Poyo variety in 1960,
- The first unsuccessful banana project by the European Development Fund (FED) in 1964,
- FED's second abortive banana project in 1967,
- The former west Cameroon penetration of the French market,
- 1968s drastic decline in small-scale production,
- Export discontinuation of the "Gros Michel" variety in September 1968,
- Initial phase of the Cameroon Banana Organisation (OCB) in 1970, and
- OCB's request for foreign aid for the establishment of a new 1,600 ha large plantation.

Table 2.7 Cooperatives and Individual Export, 1971

Names	Export (mts)
SOBACO (North, South, Central)	35,573
M. Wambo	236
M. Nuetsa	212
Coopam	123
M. Ngawa Victor	52
M. Tchuissi	17
M. Sandjo	5
Others	32

Source: Lecoq (1972), L'Evolution de l'Economie Bananiere au Cameroun. Fruits, Vol. 27, No. 10, pp. 677-696.

Chapter Three

The Banana Industry

3.1 Cameroon's Banana Industry

Cameroon's banana industry is concentrated in the Mungo and Fako Divisions of the Littoral and Southwest Provinces, respectively (Appendix III). Approximately 83% of the firms are located in the Mungo Division, along the Douala/Nkongsamba road of the Littoral Province. They are: Organisation Cameronaise de la Banane (OCB), Societe de Plantation de Nyombe-Penja (SPNP), Nassif, Plantations de Haute Penja (PHP), Tiani, Djoumo, and 40 others classified as smallholders. Smallholdings are defined as plantations between two and 50 hectares in size. OCB, a parastatal formed in 1968, operates the largest banana plantation with an area of 1,165 hectares.

Prior to 1984, approximately 17% of the banana producing firms was located in Fako Division (Ekona), Southwest Province. This plantation is owned by the Cameroon Development Corporation (CDC), which is the largest agricultural firm in the Republic of Cameroon. FAO's report on the World Banana Economy (1986) showed that the Cameroon banana industry in 1984 cultivated a total of 3,252 hectares, and exported 57,900 mts, for an average yield of 17.8 mts/ha (Table 3.1). The calculated yield per hectare is based on the total exportable output. The former West Cameroon was deprived of the British Commonwealth tariff preferences in July 1963 as she merged with the former East Cameroon on 1st October, 1961. Consequently, total banana export significantly fell to an average of 57,000 mts compared to 81,000 mts produced during 1962-65 (FAO, 1986; Nelson et al, 1974).

Epale (1985), Ngoh (1987), and Mbuagbaw et al (1987), attributed the 30% fall in total export during this period to the lost commonwealth tariff preferences. Additional losses incurred included the exchange rate parity, i.e., 692 FCFA = £1 instead of 685 FCFA, the then official rate. However, the preferences were regained in 1973 when Britain became a member of the European Economic Community (EEC).

15

Table 3.1: Cameroon: Area under bananas, exports and unit yields

Growers	Area (Ha)	Exports (1,000 mt)	Ave. Yield (mt/Ha)
Organisation Camerounaise de la Banane (OCB)	1165	20.7	17.8
Societe de Plantations de Nyombe-Penja (SPNP)	747	11.6	15.5
Cameroon Development Corporation (CDC)	566	13.5	23.9
Nassif	210	3.4	15.9
Plantations de Haute Penja (PHP)	200	3.3	16.5
Tiani	90	1.8	20.0
Others: 40 planters (2-50 ha)	274	3.6	13.1
Total	**3252**	**57.9**	**17.8**

Source: FAO (1986), The World Banana Economy.

Based on exportable yield per hectare, CDC is the best managed semi-modern plantation, followed by Tiani (Table 3.1). However, from an economic efficiency standpoint, profit per hectare is a better measure of how well a banana plantation is being managed. Since 1989, CDC has reduced its total cultivated area by approximately 50% due to several factors such as: poor quality fruits, plantation mismanagement, lack of irrigation system, high cost of production, and disease infestation (black sigatoka, nematodes and borers).

Everything being equal, CDC would be better off if it closed down its banana section. Refinancing, the hiring of qualified management and technical staff, and adopting appropriate technology would be the proper, but expensive alternative to rejuvenating the plantation. However, due to some cultural, ethnic, and social constraints, the management thinks pare-to optimality could be obtained (i.e., more social benefits) if the status quo is maintained, irrespective of the costs and negative margin constraints. Consequently, only the most unproductive areas have been eliminated.

Recently, SPNP has been engaged in a restructuring programme. The company intends to adopt modem banana technology and expand its plantation to 2,000 ha, i.e., a 168% increase. If the expansion and the technological innovation programmes are accomplished, the company is likely to become the largest in terms of size or cultivated area. Its position, in terms of yield/ha, would largely depend on farm management, adopted technology, disease control programme, irrigation and other factors. The soil in this area is better suited for banana production than in the Fako-Tiko plain (FAO, 1986). The goal of the company is to produce at least 30,000 tonnes of banana by the end of 1990 for the French market. This would be an increase of 159% compared to 1984 where only 11,600 tonnes were exported. This would require an average of 2,300 mts exported monthly compared to 892 mts in 1984. Moreover, it is estimated that 50% of the total production would be of the extra category grade, the best in the world market.

3.2 Categories of banana producers

Cameroon banana producers can be classified under four categories, namely: multinational (MNCs), parastatals, foreign, and private (Table 3.2). OCB, CDC, and IRA are owned by the Cameroon government and are heavily subsidised. SPNP and PHP are operated by French-based companies, whilst Tiani, Djoumo, and other smallholders are privately owned. Del Monte has a joint venture with CDC, and was the only multinational fresh fruit company (MNC) now based in Cameroon.

3.3 Multinationals (MNCs)

In 1987, the Del Monte Tropical Fresh Fruit Co. (DMC) contracted with CDC to open a new banana plantation in Tiko, even though CDC already had one at Ekona. The contract required CDC to provide land and labour, while DMC provided capital, production inputs, transferred technology, technical assistance, and assured quality product. The fruits produced would be bought from CDC by DMC at free-on-board (FOB) price, if the quality specifications were met.

Table 3.2: Classification of Cameroon Banana Companies

Company	Para-Statal	Private Cameroon	MNC	Foreign
OCB	×			
ODC Ekona	×			
IRA	×			
SPNP				×
PHP				×
Comp. De Bananes				×
Tiani/Djeumo/Sm.Hold.		×		
BCUF		×		
K/F of CPMS Ltd		×		
PFPC		×		
Del Monte/CDC	×		×	

Source: Fonsah, (1992), Field Survey.

DMC is one of the world's leaders in the international market for fresh fruits, especially bananas. In 1989/90 DMC supplied 966,000 mts of banana in the world market out of which 500,000 mts were exported to the United States and 250,000 mts to Europe. Out of the total export, approximately 2% came from Cameroon. The CDC/DMC Tiko banana project is targeted for 1,200 ha. Only 800 ha have already been planted while approximately 700 ha are under full production. The expansion project was expected to finish by the end of 1990. During the first semester of the 1989 fiscal year, 12,000 mts of banana were exported to France under the DMC/CDC logo. By the end of the fiscal year 1989/90, approximately 20,000 mts would be exported to France and about 500 mts to Italy and Britain. At the end of the restructurisation programme, 1990/91 yields are expected to attain 45,000 mts yearly.

The CDC/DMC plantation is highly capital intensive and semi-automated. The company has adopted the most modern management techniques and technology in banana plantation. This highly sophisticated technology can only be found in Central America and the Philippines. The increased yield of 33 mt/ha during the first year of operation is associated with its technological innovation and modem farm management strategies (Field Survey, 1990) compared to an average of 17.9 mt/ha from its competitors (FAO, 1986; Field Survey, 1990). Moreover, yield/ha is expected to increase further as production goes through a complete sequence - that is, from mother plant (madre or plantilla) to daughter plant (hijo), and subsequently to grand-daughter (nieto). Then the sequence begins again (Stover and Simmonds, 1987).

Table 3.3: OCB actual costs and returns per metric tonne.

Item	$
1. Total Revenue	484.00
2. Operating Costs	
Pre-harvest machinery fuel, lube and repair	
cost, seed, fertilizers, herbicides, labour, etc.	271.20
Harvesting and Transportation cost	
Packing cost (including cartons)	6.90
Transport to port	45.70
Loading and stevedoring	11.40
Taxes and other charges	12.60
Freight and insurance (French Port)	4.60
Handling and expenses on arrival (including	129.30
losses)	42.30
Commission (2.5% of selling price)	12.10
Total Operating cost	536.00
Net Revenue above costs	-52.10

Source: FAO (1986), The World Banana Economy.

3.4 The Parastatal Organisation

While DMC and SPNP are expanding, OCB is liquidating. For the past two decades, the OCB company has been realising negative gross margins. It has survived the past two decades because of government subsidies. With the current economic crisis plaguing Cameroon, one of the austerity measures taken by the government is to privatise most or all of the parastatal organisations. OCB is amongst those already chosen for privatisation.

The downfall of the parastatal plantations can be attributed to several factors, including mismanagement, outdated technology, lack of qualified staff, high cost of production (Table 3.3), and poor and unacceptable fruit quality which has earned Cameroon growers the lowest prices in the protected French market and loss of credibility (GIEB, 1988). Due to the negative returns, OCB orchestrated a rehabilitation program. The objective was to increase yield to approximately 80,000-90,000 mts by 1990.

3.5 Private Firms

The cultivation of export bananas has been monopolised by Europeans for so many years. It was only in 1946 that Cameroonians developed a real interest and joined the business. Since then, several individuals or group of individuals have opened banana plantations either in their own names or as cooperatives.

Some of the well known individuals in the former East Cameroon were Messrs Wambo, Nuetsa, Ngawa Victor, Tchuissi, and Sandjo. The following cooperatives were formed: SOBACO (north, south, and centre), UCAPC, COFRUICAM, UBM, COOPLABAM and UNICOPRODICAM. Currently, only Tiani is still in the banana business.

In the Southwest Province, several individual producers merged and formed cooperative unions. The most prominent ones were the Bakwerie Cooperatives Union of Farmers Ltd. (BCUF), the Muyuka Progressive Farmers' Produce Enterprise (PFPC), and the Kumba Federation of Cooperative Produce Marketing Societies (K/F of CPMS Ltd.). All these organisations were out of business by 1971. Production under these cooperatives were highly labour intensive and adopted the semi-modern technology.

3.6 Private French Firms

Banana production was started in the former East Cameroon in 1931 by French individuals and subsequently French companies. The pioneers were Messrs Adolphe Beynis and Jacqueys, who opened the first plantation at Penja. Then came Messrs Maurice Beynis and Rossides who together opened another plantation at Mbanga (Appendix III). Later on, Compagnies des Bananes, Plantation de Haute Penja (PHP) and Societe de plantation de

Nyombe Penja (SPNP) were created. All these firms had their base in France. Cameroonians were not interested until 1946. The members of the Union of Cameroon Banana Producers, Le Syndicat de Defense des Interets Bananiers du Cameroun (SDIBC) formed in 1936, were all Europeans. Until the present, Cameroon banana production is still being monopolised by the French in former East Cameroon, as the few existing smallholders have been squeezed out and the only parastatal (OCB) was bought by SPNP. The French firms are more labour intensive. They also adopt semi-modern technology (Lecoq, 1972).

3.7 Banana Marketing

Studies have shown that, compared to other fruits, such as, oranges, apples and citrus, banana is the "freshest" fruit in the market due to its highly perishable nature. That is why it is usually consumed "within three weeks after harvest". With modem technology, including harvesting practices, the consumable time could be extended to eight weeks (Badran, 1969; Deullin, 1970).

Its availability or marketability all over the world, especially to the non-producing countries, requires a well defined and orchestrated transportation network. This network includes moving the fruits from the plantation to the packing station, from the packing station to the ports of embarkation (ship), and finally to the final consumers, who buy them in any marketing outlet. The major objective of this chain of transportation network is "to deliver the fruits to the ripening room in a firm green condition as free of blemishes as possible" (Stover and Simmond, 1987).

Packing and transportation are usually the responsibility of each planter. However, the smallholders who cannot afford packing stations utilise OCB's facilities. Further, OCB assists in arranging transportation from farm to port, either by rail or truck. Del Monte, on the other hand, utilises forty-foot $(60m^3)$ containers. Initially twenty-foot $(30m^3)$ containers were also used. But during peak periods and as production increased, these were no longer needed. DMC also adopted the refrigerated containers during peak seasons and when it was anticipated that the expected weekly production capacity requirement would not be met in the given harvesting days, pre-harvesting and packing were done and the fruits were put in the refrigerated reefers against the next shipment.

After arriving at the port, the boxed bananas are loaded in the ship immediately. As soon as the bananas are loaded on board the ship, the pulp temperature is reduced. In Cameroon, the pulp temperature varies significantly between companies. It ranges between 8°C to 14°C.

Prior to 1984, Cameroon bananas were shipped at Bonaberi wharf in bunches. Today, a newly, more modern wharf with belt conveyors and cranes is located in Douala where boxed bananas are shipped. OCB is responsible for shipping arrangements based on "50:50 between Cameroon shipping lines and the French shipping company, L. Martin".

Two international shipping agencies, with representatives in Cameroon, are involved. Seatrade is represented by Cameroon Shipping Lines while Del Mas is represented by Camatrans. Cameroon growers are plagued with high shipping costs because output during the non-peak period is usually insignificant, so that chartered ship companies refuse to supply their vessels.

Cameroon bananas are loaded in non-container ships. Loading is usually done by belt conveyors and cranes (on semi-pallets platforms) simultaneously. Exported bananas vary in box weight. DMC and SPNP bananas are packed in 20.65 kg boxes while OCB and CDC utilise 17 kg boxes.

The number of boxes to be stacked (loading plan) varies from company to company, and depends on many factors, namely, quality (mt) to be shipped, size of the hatch, etc. DMC accepts a maximum height of 8-9 boxes while the other growers accept between 8-13. The number of boxes to be stacked at a given tune has a lot of impact on the quality of the fruits. Loading is done continuously for a maximum of three days or 72 hours. It is the growers' responsibility to take their fruits to the port on time.

OCB has a very important role in the French market. It represents, or rather plays, the role of a sales agent/distributor of Cameroon bananas to the French commission agents for a commission fee of "2 1/2% of the sales value in 1984". Further, a quota is set where each commission agent is told the quantity to be supplied at each given time. Recently, Simba France has joined the number of commission agents dealing with Cameroon bananas. However, Simba France is the sole distributor of DMC bananas in the French market. Other agents include: Compagnies des Bananes (United Brands), Agrisol, Compagnie Fruitiere, Siim, Muribane, and Ste Dunand et Cie (FAO, 1986).

3.8 Domestic Banana Marketing

There has always been a domestic market for bananas even before 1907, when the first banana plantation was established. Only the local variety was cultivated and sold. The initial banana shipment to Germany was also the local variety that was dehydrated. At that time, the fruits were sold in bunches.

Contemporaneously, the local market for bananas has developed tremendously. Improved road infrastructures have facilitated the transportation of goods from areas of abundant supply to areas of scarcity, such as from the Southwest or Littoral Provinces to the Northern or extreme North Provinces, where banana cultivation is non-existent. Further, the entrance of several firms into the export banana business has increased the supply available for the local markets as all their rejects are sold locally.

Domestic prices vary from province to province and from market to market. In 1980, a report by the Republic of Cameroon, Annuaires Statistiques (1983) showed that the highest price/kg of banana was 99 FCFA in Garoua in the Northern Province, while the least was 24 FCFA at Bafoussam in the Western Province. Recent studies by Almy et al (1990) showed a two years' average of 38 FCFA/kg for banana in the Southwest Province. The lowest average of 25 FCFA/kg was recorded in Meme Division, while the highest, of 45 FCFA/kg, was in Ndian Division.

3.9 The Nutritional and Therapeutic Importance of Bananas

The banana has multifaceted importance. In Africa, Central America, the Caribbean Islands and some Asian countries, banana is a staple food. It is cooked green, fried, or eaten ripe (Stover and Simmonds, 1987). In some East African countries, such as Uganda, Rwanda, Burundi, Zaire and Northern Tanzania, ripe banana is used to make beer and wine (Tibaijuka, 1985; Acland, 1971). The sugar content in the banana peel of the "valery" variety is suitable for "yeast fermentation" (Goewert and Nicholas, 1980). Ripe banana can be used to produce vinegar (Adames, 1978), and alcohol if combined with other compounds (Maldonado et al, 1975). Studies by Chung and Meyers (1979) revealed that banana waste and pulp can be used to produce bioprotein.

Bananas can be processed for foods, such as, puree, canned slices, flour, frozen, chips, essence powder, juice, jams, flakes, freeze-dried slices in vinegar, alcoholic beverages, and a "filler" in catsup. It can be used to complement dairy products, such as, ice cream or yoghurts. It can equally be used for bakery and food processing, such as, banana bread, cake, cream pie, muffins, baby food drinks, fruit drinks, cereals, and banana apple sauce (Wilson, 1975; Stover and Simmonds, 1987; Crowther, 1979; Dupaigne 1967; Viquez et al. 1981; and Sanchez-Nieva et al. 1975). Furthermore, the banana pseudostem was used to produce starch (Simmonds, 1966; Shantha and Siddappa, 1970; Ffoulkes et al, 1978; Kayisu et al, 1981; Lii et al, 1982). According to Ffoulkes et al (1978), banana leaves and pseudostems can be a good source of animal feed during dry spells when vegetation is limited.

On the other hand, the therapeutic importance of the banana includes: low fat, cholesterol, sodium and salt content, and rich in potassium, approximately 400 mg/l00g pulp (Stover and Simmonds, 1987). Gasster (1963) recommended banana fruit to corpulent and elderly patients due to its low fat elements and superior energy content. Mitchell et al, (1968) suggested it to individuals afflicted with poor digestive secretion problems. Koszler (1959) approved it for children suffering from induced or infectious abdominal malaise, while Seelig (1969) endorsed it for the treatment of nutritional imbalance and intestinal inflammation in children. Furthermore, banana contains "ascorbic acid and vitamin B", especially the "Gros Michel" variety (USDA, 1963; Stover and Simmonds, 1987).

Part Two

Theoretical Issues in Banana Production and Marketing

Chapter Four

The Supply of Bananas

Introduction

This section deals with the production (supply) and marketing (demand) of Cameroon bananas. The supply-side focuses on the economic aspect of banana production, dwelling on the several inputs and a single output model. A banana production function (supply) contains a combination of both fixed and variable inputs. Indeed, it would be difficult to measure these inputs in physical terms. As a result, only the variable inputs, such as, fertilisers, were treated (Debertin, 1986; Beattie and Taylor, 1985; Silberberg, 1978).

On the marketing (demand-side), a simple adjusted international trade model was adopted (Section 7.1). Expansion of this model has been done by numerous international trade theorists in various agricultural commodity world markets, such as, food grains (e.g. wheat and rice), feed grains (e.g. corn barley and oats), and seed crops (e.g. soybeans and sunflowers) (Kost, 1976; Wilson and Takacs, 1978; Collin et al, 1980; Goldstein and Khan, 1976; Learner and Stern, 1970; Chamber and Just, 1978).

One of the specific objectives of this study is to compare banana production under traditional, semi-modern and modern technologies. These technologies are discussed with the aim of highlighting their differences and similarities. Furthermore, yield and price trends in both the open and preferential markets, seasonal price movement, marketing costs and marketing structure in the domestic marketing were all highlighted to meet with the other objectives.

4.1 Banana Production Theory (supply)

Banana production depends on the following major factors represented in the equation (4.1) all things being equal:

$$S_{cb} = g \ (S_{mt}, F_{It}, P_{dc}, I_{rr}, ..., Z_n)$$
where:
S_{cb} = The Supply of Cameroon Bananas (output);

S_{mt} = Soil management;

F_{it} = Fertility program;

P_{dc} = Pest and disease control program;

I_{rr} = Irrigation network;

Z_n = Other factors, such as weather and adopted technology.

4.1.1 Soil Management (S_{mt})

Soil generally contains minerals, organic matter, water and air. According to Jones (1982), 50% of the soil volume is mineral and consists of 93% silica (SiO_2), aluminium and iron (Al_2O_3 and $_{Fe2}O_3$), respectively. Calcium (Ca), magnesium (Mg) and potassium (K) oxides together make up four percent, while titanium (Ti), sodium (Na), including an infinitesimal quantity of nitrogen (N), sulphur (S), phosphorous (P), boron (B), manganese (Mn), zinc (Zn), copper (Cu), molybdenum (Mo), chlorine (Cl), etc., make up only three percent (Table 4.1).

Table 4.1 Chemical Composition of a Mineral Topsoil Expressed as Oxides of the Elements

Oxides	Percent
SiO_2	76
Al_2O_3	12
Fe_2O_3	5
CaO	1
MgO	1
K_2O	2
All others	3

Source: Jones U.S. (1982), Fertilizers and Soil Fertility, 2nd Edition, New Delhi: Prentice-Hall of India Private Ltd.

Jones (1982) further explained that out of the three highest ranking chemical properties depicted in Table 4.1, i.e. SiO_2, Al_2O_3, and Fe_2O_3, only a negligible quantity of iron (Fe_2O_3) is required for "plant growth". Studies by Walmsley and Twyford (1976); Lahav and Turner (1989) confirmed that a normal plant's maximum uptake rate is 1gr of iron (Fe) in which 80% occurs at the initial stage. However, huge amounts of SiO_2, Al_2O_3 and Fe_2O_3 are required to support the root system, the colloidal clay, and humus. Table 4.2 depicts the essential plant nutrient elements.

Table 4.2 Essential Plant Nutrient Elements

From Air and Water	Macro-nutrients From Soil Solid and other chemicals	Toxic Element	Micro-nutrients From Soil Solid and other chemicals
Carbon	Nitrogen	Selenium	Iron
Hydrogen	Phosphorus	Arsenic	Manganese
Oxygen	Potassium	Molybdenum	Zinc
	Sulphur	Fluorine	Copper
	Calcium	Aluminium	Chlorinie
	Magnesium	Nickel	Boron
			Molybdenum
			Cobalt

Source: Jones U.S. (1982). Fertilizers and Soil Fertility, Prentice-Hall of India, New Delhi. Wrigley G. (1902). Tropical Agriculture, ELBS/Longman, 4th Ed., p.21.

A productive soil has a good water-holding capacity, sufficient aeration, and a supply of decomposing minerals and humus that are dissolving at a rate rapid enough to meet the requirements of desired plant growth" (Jones, 1982; Stover and Simmonds, 1987; Lahav and Turner, 1989). Consequently, obtaining the maximum yield and best quality of any plant (bananas in this case) largely depends on how well the soil is managed. Early identification of nutrient deficiencies and the recommendation of adequate drainage systems are partial solutions to maintaining the quality of the soil (Lahav and Turner, 1989). Other conditions include soil characteristics, texture, pH level, Cation Exchange Capacity (CEC), and the structure.

The Cation Exchange Capacity (CEC) is an important factor in determining the quantity of various nutrients required by the plant. An elevated CEC soil, for instance, indicates the need for additional potassium (K^+), and vice versa. If this need is not satisfied (potassium deficiency), it becomes a major problem (Jones, 1982).

The analysis of soil acidity and alkalinity or pH (-log of the effective hydrogen ion concentration) plays a vital role in banana production. The soil pH ranges between 1 and 14. If the pH is less than seven, the soil is classified as acidic. However, a pH greater than seven means that the soil is alkaline, whereas, a pH of seven symbolises neutrality. (Cooke, 1982; Jones, 1982; Clarke, 1988).

Lahav and Turner (1989) pointed out that successful banana cultivation can be undertaken in soils with pH 3.5 to 9.0. However, they added that the soil pH of 5.5 to 8.0 is ideal. Champion et al (1958) experienced a better banana yield in Guinea planted on soils with pH 6.0 than on soils with pH 4.5. The pH of 5.5 to 6.8 is ideal in Cameroon.

Godefroy et al (1978) obtained a "positive correlation" in their studies of the effect of lime application and the response to banana yield on soils with pH varying from 3.4 to 6.7 in Cote d'Ivoire (Table 4.3). Studies by Garcia et al (1976) showed a negative correlation in the size of the pseudostem when planted on soils with pH ranging from 4.4 to 8.2 in the Canary Islands.

Table 4.3: The effect of lime application to a peat soil on pH and banana yields in Cote d'Ivoire

Treatment (ton/ha)	pH[2]	Yield (t/ha) crop cycle			
		1	2	3	Total
0	3.4	43.8	48	39	130.8
3	4.2	43.4	47.9	41.6	132.9
6	4.7	43.4	47.8	42.3	133.5
12	5.9	46.2	48.6	44.1	138.9
24	6.7	44.3	49.5	43.3	137.1
		NS[3]	NS	NS	

Source: Lahav and Turner (1989), Fertilizing for High Yield Banana, IPI Bulletin 7, 2nd Ed., Berne/Switzerland, p 9.

[2] after two years. [3] no significant difference.

4.1.2 Soil fertility (F_{1t})

Although most of the nutrients listed in Table 4.2 are necessary for banana production, nitrogen (N), potassium (K) and phosphorus (P) are the most important, followed by sulphur (S), calcium (Ca), and magnesium (Mg). These six nutrient elements are called "macronutrients" (Jones, 1982; Cooke, 1982; Wrigley, 1981). Soil fertility analysis will be strictly limited to only the macronutrients in this study.

A banana plant consumes an amazing quantity of mineral nutrients which are frequently in limited supply. The soil of a well managed plantation producing 50 t/ha/yr loses approximately 1,500 kg N/ha/yr. Consequently, replenishment of the lost nutrient becomes a necessity in order to maintain not only the soil fertility but also a constant yield pattern. In the traditional cultivation system the application of organic manure suffices to make up for the lost nutrients, whereas the semi-modern and modern production systems require fertiliser application (Lahav and Turner, 1989).

Furthermore, Lacoeuilhe and Martin-Prevel (1971), Marchal et al (1972), Marchal and Martin-Prevel (1971), and Warner and Fox (1977) recommended "plant and soil" analysis as a significant field experiment necessary to determine the quantity of fertiliser needed to maximise output. Table 4.4 and Appendix V give a breakdown of symptoms on leaves and fruits associated with various macro and micronutrients deficiencies in banana production.

Table 4.4: Summary of Excess Symptoms[4]

Symptoms on:	Descriptions of symptoms	Element
Petioles	Blue	Mg
Leaf	Irregular chlorosis followed by necrosis	Mg
	Marginal chlorosis followed by necrosis	Na, B
	marginal blackening followed by necrosis	Fe, Mn
	Chlorotic striping	As
Fruit	Not filled	As
	Not filled	Cl
	Weak bunch, widely spaced hands	N
Roots	Growth inhibited	Cu

Source: Lahave E. and D.W. turner (1989), Fertilising for High Yield Banana, 2nd Ed., International Potash Institute, Bulletin 7, Berne/Switzerland, p. 13.

4.1.2.1 Nitrogen (N)

Butler's (1960) fertiliser experiments in Central America with the Gros Michel banana variety depicted nitrogen (N) as the most significant element needed for banana production. Despite the fertility level of the soil, N is always inadequate. Twyford (1967) and Jones (1982) ranked it the second most needed element for banana and non-leguminous plant cultivation after potassium (K). Most of the deficiency symptoms cited by Murray (1959) correspond with those listed in Tables 4.4 and Appendix V respectively. These symptoms are pretty obvious if the following conditions exist: (a) inadequate drainage; (b) insufficient water (dry spell); (c) unproductive root system; and (d) lack of weed control. However, nitrogen deficiency can be corrected by a simple application of any of the existing fertilisers such as a urea, NH_4NO_3, $(NH_4)_2SO_4$, etc. Twyford (1967) emphasised that the "form" utilised must be a function of the soil pH and water availability. Lahav et al (1978) recommended potassium nitrate fertiliser for banana, due to its dual function. Due to cost constraints, only few banana growers utilise this fertiliser.

[4]Also see Ben Meir (1979); Charpentier and Martin-Prevel (1965); Fergus (1955); OsaLatz (1952); and Stover (19720.

Generally, nitrogen application rate and frequency vary from country to country and from plantation to plantation. However, the range varies from 250 to 600 kg N/ha/yr, four to six times yearly.

According to Jones (1982) urea is the most popular solid nitrogen fertiliser currently available in the international market even though Twyford (1967) disapproves of its use during dry spells except when irrigation is available. A sound knowledge of the plant N requirement and possible inhibiting factors such as the weather is advantageous.

4.1.2.2 Potassium (K)

As earlier mentioned, Twyford (1967) and Jones (1982) rank potassium (K) as the most significant macronutrient element for banana cultivation. Fourcroy and Vauquelin (1807) were the first to depict the increased level of K-content in banana. For the past century, researchers in various parts of the world have arrived at the same conclusion. Magee and Fitzpatrick (1932), Simmonds (1966), Missingham (1962) and Brady (1974), just to name a few, enumerated some common K deficiency symptoms which include among others: leaves dropping, leaves turning yellow too early, and/or "the appearance of orange-yellow chlorosis of the oldest leaves and their subsequent rapid death". Furthermore, and as pointed out by Lahav (1972), Martin-Prevel and Charpentier (1963), and Murray (1959), lack of K can cause the plant to choke, decrease the size of the leaves, increase flowering time, and decrease bunch weight, size and hand class. These pomological factors are very crucial to commercial banana firms and are essential in projecting future harvesting.

Timing and a broad knowledge of banana K uptake behaviour in the plantation is absolutely necessary. For example, Martin-Prevel (1967) revealed that K-uptake in the banana plant is highest at the initial stage of planting. Consequently, better results could be obtained if the K application is adopted during the early plant crop phase, whether or not the quantity supplied is adequate. However, and as pointed out by Twyford and Walmsley (1974), a large amount of K is also needed during the later phase of the plant crop life span. On the other hand, a lesser quantity is needed during the peeping/bending stage of the inflorescence, while die need for K in the ratoon (daughter plant) is insignificant. Studies by Vadivel and Shanmugavelu (1978) showed that an adequate application of potassium fertiliser not only improves productivity in terms of total output but also improves the general quality of the bunch. The additional utilisation of K increases the "sugar/acid ratio" which subsequently decreases the acidic level of the fruit. The following were recommended for banana cultivation: (a) potassium chloride, (b) potassium sulphate (K_2SO_4), and (c) potassium nitrate (Lahav and Turner, 1989). Studies from other countries indicate different application rates ranging from 300 to 1300 Kg/Ha/yr.

33

4.1.2.3 Phosphorus (P)

Phosphorus is a very important macronutrient that stimulates the evolution of plants in the field. P-deficiency may have the following dual repercussions: (a) reduce growth rate, and (b) deprive the uptake of other required nutrients (Brady, 1974; Wrigley, 1981).

Studies by Croucher and Mitchell (1940) and Walmsley et al (1971) showed that very little quantity of phosphorus is needed by the banana plant. Martin-Prevel (1978) supported the above assertion but explained that the reason for the low P-uptake is due to the migratory nature of the element and the ability of the plant to re-use the available phosphorus. Consequently, P-leaching has no immediate negative effect on plants. Since drip irrigation induces P-leaching, Lahav (1977) and Lahav et al. (1978) recommended the use of P-fertiliser from time to time.

Even though deficiency symptoms are seldom identified at the plantation, it has an adverse effect on total output. Some of the symptoms enumerated by Lacoeuilhe and Godefroy (1971), Simmonds (1966), and Charpentier et al (1965) include "stunted growth" and unsatisfactory root evolvement. Others have been cited in Table 4.4. Slightly similar to potassium, P-uptake in tropical conditions is highest during the mid-plant crop phase, i.e. between 8 to 20 weeks and declines by 20% after the baggable stage (Walmsley and Twyford, 1968). Lahav (1989), and Lahav and Turner (1989) cited super phosphate and rock phosphate as the popular brand of P used in bananas and recommended 100 kg/ha/yr.

4.1.2.4 Calcium (Ca)

According to Cooke (1982), calcium plays two important roles. First, it is a vital nutrient even though very little quantity is needed by most plants. Secondly, calcium "dominates exchangeable cations" by neutralising the soil's reaction. For instance, if the soil contains excessive quantities of calcium, the "negative charges" are neutralised by Ca_2+ ions. On the other hand, an inadequate soil calcium content caused by leaching is rendered acid by the "positively charged hydrogen ions". Consequently, liming becomes necessary even though not usually regarded as fertiliser.

Studies of the symptoms of Ca deficiency in bananas conducted by Freiberg and Stewart (1960), Murray (1959), Cann (1964), Martin-Prevel and

Montagut (1966), and Charpentier and Martin-Prevel (1965), include "sawtooth" semblance, and "spike leaf" emergence (Appendix V). However, other factors such as frost and accelerated development have been associated with the "spike leaf" syndrome. Nevertheless, Charpentier and Martin-Prevel (1965) associated the ripe banana cracking peel and poor appearance to Ca deficiency. For banana, super phosphate (21% Ca) approximately 3-6 t/ha, for a period of 3-5 years was recommended. Calcium can also be supplied as lime or dolomite (Lahav and Turner, 1989).

4.1.2.5 Magnesium (Mg)

According to Cooke (1982), Mg deficiency is easily detected on plant leaves such as potatoes, tomatoes, sugar beets, etc., but seldom have a negative effect on yields. "Light soils" in heavy rainfall areas are more susceptible to Mg deficiency than heavy soils whereas sandy soil naturally hold very little magnesium.

"Dolomitic limestone" is the easiest and most economically efficient method of preventing Mg deficiency on low pH soils, whereas magnesium salt is preferable on soils with pH 7. The abundant application of fertilisers usually reduce the pH of light soils due to lost Calcium.

Any rectification, liming for instance, that ignores the roles of Mg, creates further disturbances due to Ca/Mg unbalance. Consequently, Mg-deficiency becomes obvious. Brady (1974) cited "erosion, crop removal and leaching" as the three ways in which Mg can be lost from soils.

Lahav and Turner (1989) also pointed out the effect of Mg-deficiency on bananas (Table 4.4 and Appendix V). Chalker and Turner (1969) added that Mg deficiency becomes obvious in banana plantations only after cultivating for 10-20 years without Mg application. Messing (1974) added that Mg deficiency in bananas is also possible where an increasing quantity of K fertiliser application occurred for several years.

Charpentier and Martin-Pre'vel (1965) showed that bananas cannot survive if there is absolutely no quantity of Mg in the soil. The suitable form of Magnesium sulphate ($MgSO_4$) which is applicable to the soil and the spray form, applicable only to the banana leaves, were recommended for field use. The following were recommended for field use: (a) $MgSO_4$, soluble form, and (b) $MgSO_4$, the spray form applicable only to the banana leaves (Mirza and Khalidy, 1964).

4.1.2.6 Sulphur (S)

Although a macronutrient, most tropical soils are sulphur deficient. Even though coal and oil are natural sources of this element, plants in the tropics do not benefit from these sources due to less industrialisation as the quantity of S required may be obtained from air or soil (Brady, 1974; Cooke, 1982). Another source of sulphur is the soil organic matter which has to become inorganic before S is available to the plants. The complete process is rather nonchalant and sometimes non-existent, depending on whether the transformation from organic to inorganic form occurs or not (Wrigley, 1981; Cooke, 1982).

Studies in England show that approximately 20.5 kg/ha of sulphur can be obtained from rainfall. Plants grown in the outskirts of industrial cities can obtain an average of 30% of their sulphur requirement from the atmosphere. However, when intensive cultivation is undertaken, supplementary S in the form of fertilisers such as gypsum ($CaSO_4$, $2H_2O$) is required to reinforce the inadequate natural source (Brady, 1972; Cooke, 1982). Several methods of eliminating S deficiency can be adopted if inexpensive.

Walmsley and Twyford (1976) reported that the banana absorbs more S from planting to the inflorescent stage. Thereafter, the rate declines. The S required for the bunch development is extracted directly from the leaves and pseudostem. The following were recommended for bananas: $(NH_4)_2SO_4$, K_2SO_4 or super phosphate. In Cameroon, Marchal et al (1972) experienced a slight improvement in yields with S application at the rate of It/ha/yr, only in the ratoon. S deficiency is similar to calcium (Ca) and Boron (B), respectively. Some deficiency symptoms are listed in Appendix V.

Chapter Five

Pest and Disease Control

5.1 Sigatoka disease

According to Meredith (1970), Stover and Simmonds (1987), Foure and Lescot (1988), *Sigatoka* was first identified in Java in 1902. Since then, the disease has spread to almost all banana and plantain producing countries around the world. For instance, it was reported in East Africa in 1938 and subsequently in West Africa in 1940. Further incidences were reported in Honduras in 1972 and Zambia in 1973 (Stover 1980). It was first reported in Gabon and Cameroon in 1979 and 1983, respectively. In 1986, it was reported in Ivory Coast, Ghana and Nigeria (Inibap Annual Report, 1987; Foure, 1982, 1984, 1988; Stover and Simmonds, 1987). However, and according to Lecoq (1972) the *Sigatoka* disease has been present in the Muyuka area of the Southwest Province, Cameroon, since 1950. A control system using aerial spray started in 1957.

Further, Inibap Annual Report of 1987 attributed the 30-50% decrease in banana/plantain production in Gabon and Cameroon to the *sigatoka pathogen*. The disease is more devastating in modern and semi-modern plantations for the following reasons:

(a) Their products are strictly for the export markets with well-defined quality specifications;

(b) It causes premature ripening between the port of embarkation and final destination, resulting in fruit rejection and/or heavy losses;

(c) It renders the leaves non-functional as they grow older prematurely, forcing excessive deleafing which increases production costs as additional manpower and/or overtime will be required for the operation and thus depresses yield;

(d) As a result of (c), premature harvesting is induced in infected areas (heavy loss in terms of calibration and weight);

(e) Also as a result of (c), "photosynthetic leaf surface" is reduced; and

37

(f) It largely reduces profitability by more than 30% due to the loss in weight/bunch/ha, and as a result of high cost of control (Four6 and Lescot, 1988; Stover and Simmonds, 1987).

Studies in Central America showed that the cost of controlling *Mycosphaerella fijiensis* has more than quadrupled as 35 to 45 spray cycles per year are usually required. Furthermore, experience in Honduras in 1984 showed that 26% production cost was generated from Sigatoka Control (Stover and Simmonds. 1987). In Guadeloupe, the number of sprays declined fourfold, from 25 to 6 while in Cameroon, an average of 10 to 12 annual sprays are commonly practised (Foure, 1986).

Berg (1980) pointed out that, approximately, 4.5 kg per bunch of bananas is lost in a highly black Sigatoka infested area. Field experiments conducted at Nyombe, Cameroon in 1989 by the Institute of Agronomic Research (IRA) showed that a bunch of plantain treated against black Sigatoka disease weighed more than a relatively untreated bunch by an average of 4.2 kg. Consequently, the Sigatoka pathogen has the following repercussions to the traditional farming system:

(a) Banana/plantain leaves are rendered non-functional due to premature ageing which depresses yields;

(b) It restricts "photosynthetic leaf surface"; and,

(c) It reduces net returns by more than 30% due to the loss in weight/bunch/ha (IRA Nyombe Cameroon, 1988/1989 Annual Report.)

Three causative agents have been identified with the Sigatoka disease. *Mycosphaerella musicola,* for yellow Sigatoka, *Mycosphaerella fijiensis* for black leaf streak, and *Mycosphaerella fijiensis. var. difformis* for black Sigatoka. Because of the close similarities (except in the sizes of the sporodochia found in the immature lesion) of M. *fijiensis var. difformis,* most authors identify them as M. *fijiensis.* (Leach 1946; Meredith, 1970; Stover, 1980; Stover and Simmonds, 1987; Foure, 1988; Foure and Lescot, 1988).

Extensive studies carried out by Foure in Gabon in 1982 and in Cameroon in 1984 depicted *Mycosphaerella musicola* as the main cause of yellow Sigatoka in bananas. On the other hand, *Mycosphaerella fijiensis* attacks both bananas and plantains. *M. fijiensis* was first discovered at Kribi along the Cameroon/Guinea/Gabon border in 1980. In 1983, it appeared in the Mungo, Littoral province. Currently, it is present at Ekona, Tiko and Buea in the South West Province and also in Yaounde in the Central Province. Further studies by Foure and Lescot (1988) depicted that when both pathogens are present, i.e., *M. mmicola* and *M. fijiensis,* the latter dominates.

The development of the *Sigatoka pathogen* is highly influenced by heavy sporadic rainfall which creates a favourable condition for the "release of ascospores" (Foure, 1986) and "perithecial production" (Meredith., 1970). The development of spotting is influenced by the climate and is most favoured between 21 °C to 28 °C (Stover and Simmonds, 1987). *Mycosphaerella fijiensis* produces lesser conidiophores than M. *Musicola* (Foure, 1986). Furthermore, the only dissimilarity between *M. diformis,* the causal agent of black Sigatoka, and M. *fijiensis var. diformis,* the causal agent of black leaf streak, is that the immature lesion of the former contains diminutive sporodochia compared to the latter (Stover, 1980; Stover and Simmonds, 1987). The conidiophores and ascospores of this inoculum are spread to other plants either by water or wind (Stover, 1980; Stover and Simmonds, 1987; Foure, 1986).

5.2   Fusarium   Oxysporum   F.sp.   cubense   (Panama   disease)

Panama disease *(Fusarium oxysporum f. sp. cubense)* is the most destructive banana diseases. According to Lecoq (1972), the disease first appeared in Cameroon around the Tiko area of the Southwest province in 1941. Mass destruction was reported seven years later and by 1954, production declined by 30% as a result of the disease. Further reports indicated that the disease spread all over the zone except the northern part of Tombel, and was very susceptible to the "Gros Michel" variety. The disease is no longer a threat since the adoption of the Cavendish varieties, but it is still a threat in other parts of the world. (Hwang and Ko, 1986; Pegg and Langdon, 1986; Deacon, 1984).

The disease affects the banana plants through the main and lateral roots system. Control methods include either "quarantine" or the use of a resistant variety (Stover, and Simmonds, 1987) but complete eradication has been futile (Hwang and Ko, 1986).

5.3 Banana Bunch-Top Virus (BBTV)

Banana Bunch-Top Virus (BBTV) is the most dangerous disease affecting bananas (Dale et al., 1986; Dale, 1987). The disease is caused by the *aphid Pentalonia nigronervosa* (Magee, 1927, 1936; Hutson and Park, 1930). It was first identified in Fiji in 1889 even though its existence went way back to 1879 (Simmonds, 1966; Dale, 1986) It was reported in Taiwan and Egypt in 1901, in Sri Lanka and Australia in 1913 (Dale, 1986), in Burundi, Rwanda, Zaire and Gabon in 1982 (Foure and Mansen, 1982; Tezenas du Montcel, 1987; and Stover and Simmonds, 1987), and in China (Liang Li-Feng, 1989).

The disease is also present in the Philippines, Western Samoa, India and Benin Islands. It has yet to penetrate Central and South America and the Caribbean (Wardlaw, 1961; Dale, 1986; Stover and Simmonds, 1987). The virus transmission is possible at a limited range of eighty-six metres. Distances beyond eighty-six metres are conceivable only if caused by wind (Allen, 1978). There is a high probability that the BBTV will eventually reach Cameroon since it is already present in Gabon. Banana producers and researchers should therefore be prepared for any possible attack.

The BBTV virus is the most dangerous because it attacks all the developing suckers from the host mother plant and deprives bunch emergence. This virus can also cause the mother plant to bear very poor quality fruit thus reflecting a great loss to the entire production unit (Simmonds, 1966). Despite Jose's (1981) findings, all the Musa spp are susceptible to the virus and all the selection and breeding strategies to develop a resistant variety have thus far been abortive. The only hope is the adoption of the genetic engineering technique (Dale, 1986). If not controlled, the disease is capable of destroying a whole plantation (Stover, 1972).

5.4: Cigar-End Rot

As the name implies, this fungus attacks the emerging bunch. The tip rots and looks like a burning cigar. It is caused by a parasite, *Trachysphaera fructigena*. In Cameroon, it affects plantations located on higher elevations and low temperatures such as the Molyko, Muyuka and Tombel areas. It is also present in the Tiko plane. In 1955, 200,000 bunches were rejected from plantations in the Southwest province of Cameroon due to Cigar-end rot (Lecoq, 1972).

Control measures include, early bagging, field deflowering technique, and fungicide application. Even though the application of Thiabendazole and Benzimidazole were ineffective (Lecoq, 1972), "1,000 ppm of Ridomil" - sprayed at two intervals - was recommended by Tezenas du Montcel (1972).

5.5: Moko Disease

Moko is a serious banana disease caused by *Pseudomonas solanacearum*. According to Simmonds (1966), Tezenas du Montcel (1987), Reville and Vargas (1967), Stover (1972) and Coonshaw and Edmunds (1980), the spread is limited only to Central and South America. However, Gnanamanickam et al (1979) revealed its presence in South India and the Philippines. Thus far, no serious damage has been reported from Cameroon or other African countries.

It is usually difficult to differentiate Moko from Panama disease due to the close similarities of the symptoms (which include the falling of both immature and mature leaves, growth depression of smaller suckers, and development of ebony and/or jaundice-like colours, interior discoloration, chocolate dry-rot like pulp, and an occasional ripe finger in between an unripened bunch (Wardlaw, 1972; Buddenhagen, 1960, 1961; Simmonds, 1966; Feakin, 1971; Stover and Simmonds, 1987).

Various studies reveal that early identification of the disease and immediate elimination of the infected areas are the best and economically efficient control strategy. Elimination requires the use of well trained agricultural technicians and the sterilisation of all equipment used. Chemical elimination of unwanted suckers is also advised to prevent the spread of the disease. Further, unproved drainage, increased pruning frequencies, rampant cross-examination (Feakin, 1971) and weed control (Sequeira, 1962; Stover, 1972) can considerably minimise infestation.

5.6: Blood Disease

The disease invades the root system of non-resistant varieties causing heavy damage. The name is inherited from the reddish fluid that comes out of the pseudostem when chopped down (Subijanto, 1990).

Although it has not yet appeared in Africa at large and Cameroon in particular, the blood disease is one of the major threats to banana production. It is provoked by the bacteria *Pseudomonas celebensis*. It has already been reported in South Sulawesi, Indonesia, and West Java.

5.7: The Banana Weevil Borer

This insect pest attacks and causes tremendous damage to the root system and the rhizome of the plants. Further, the female adult weevil inserts holes in the rhizome for larval feeding, thus exterminating the vascular tissue, which also reduces water and nutrient uptake. As a result, the plant tipover rate increases. Younger plants die, and mature plants produce inferior quality bunches, thus causing a significant loss to the firm or the individual producer (Tezenas du Montcel, 1987).

Even though it originated from South-East Asia, banana borer, *Cosmopolites sordidus,* is now present in all the banana producing countries of the world (Cuille, 1950; Simmonds, 1966).

Simmonds (1966) recommended four control methods: cultural, biological, trapping, and insecticides. The cultural method includes destroying all the pseudostems after harvest, improving weed control, and improving the selection of planting materials by peeling and dipping it into hot water (about 55 °C) for at least ten minutes (Sarah, 1990a; Pinochet, 1988a). The adoption of predators was recommended for biological control but Ostmark (1974) disapproved of the measure as the banana weevil borer has very limited "natural enemies". Thus far, biological control method has been abortive in Africa, South America, and Australia (Neuenschwander, 1988) except in Cuba where Roche and Abreu (1983) successfully reduced the population by 80% using ants, *Tetramorium guinieensis,* as predators.

Two trapping methods - the "pseudostem traps" and the "disc-on-stump trap" - have been widely recommended (Hord and Flippin, 1956). However, the trapping techniques partially reveal the existence of the pest but hardly indicate the level of damage already done. Insecticide application is effective but very expensive. Effectiveness also depends on the types used. Temik was more effective than Furadan, Counter and Nemacur in field trials in Cameroon, but Temik has been banned. The level of damage is more pronounced in the traditional production system as absolutely no control method is implemented.

5.8: Nematodes

The burrowing nematodes are as dangerous to bananas and plantains as the black Sigatoka pathogen. Several species exist and are widespread all over the world. The *Pratylenctudae, Radopholus similis, Pratylenchus coffeae, P. goideyi,* and *Helicotylenchus multicinctus* exemplify some of the species. It is widely accepted that *Radopholus similis* is the most dangerous species to banana and plantain (Sarah, 1990b; Tarte et al, 1981; Pinochet, 1988b; Blake, 1972). After nematodes attack the roots and/or rhizomes, the roots rot due to the brandy-red lesion and are rendered vulnerable for bacterial, fungi, and worm infection. As a result, frequent uprooting occurs especially when the plants are not well supported. A slight wind can destroy the whole plantation if heavily infested (Pinochet and Stover, 1980; Stover and Simmonds, 1987; Pinochet 1988b).

Gowen (1979), and Pinochet (1988b) recommended propping and guying, nematicide application, and the choice of resistant varieties as the best control technique. Propping and nematicide application have been adopted with considerable success in Cameroon. The field trials and chemicals were for both borers and nematodes. Effectiveness is highly correlated with application frequencies.

5.9: Land Snails

Land snails are as devastating as nematodes if not worse, because they feed on the leaves and fruits of several kinds of plants. Various species have been widespread all over the world either intentionally (e.g. as a food source or as a collectors item), or unintentionally (e.g. through planting material) (Mead and Paley, 1992; Dakle, 1969; Burch, 1960). Some of the well-known species include the European brown snail *(Helix aspersa),* the *Otala lactea,* the giant African snail *(Achatina fulica* and *Limicolaria aurora),* the *Veronicella leydigi* and the *Hawaii minuscula* (Bruch, 1960; Dekle, 1969; Mead and Paley, 1992).

Spence (1938) revealed various attacks by *Limicolaria aurora,* (one of the snail species), on palm fruits and other vegetative plants in Cameroon. Frankiel (1989) also disclosed the presence of another species, the giant African snail, *Achatina fulica,* in Guadeloupe, the French West Indies, and various parts of the Island. Adamu and Fonsah (1992) revealed serious raids by *Limicolaria aurora* snails and other smaller species on both traditional and commercial banana plantations in Cameroon. The smaller species were more threatening to banana plants due to their ascending flexibility.

Snail assaults differ from nematode or borer, as their primary target is the bunch rather than the roots or rhizomes. Snails prefer group assault, and when successful, the bunch is completely disfigured. This characteristic coupled with its exponential multiplicative nature ranks the land snail the most dangerous pest in export banana production (Gunn, 1924; Basinger, 1931; Burch, 1960) where quality standard is the necessary and sufficient condition for competitiveness. It is capable of exterminating an established plantation in a short time simply by eating up the fruits. Studies on land snails carried out in the only existing modem plantation by Adamu and Fonsah (1992) also confirmed that a combination of physical, mechanical, and chemical control devices are required.

5.10: Peel-feeding Caterpillar

Peel-feeding caterpillars are potential threats to commercial banana production in Cameroon. Currently, caterpillar damage has not reached economic level. Empirical studies indicate that sooner or later population explosion of this pest may occur. If it does, it will be a major blow to producers as the control measures are not yet determined in Cameroon. However, studies on the behaviour, characteristics, and species identification of the caterpillars are being conducted in the only modern plantation under study. The damage done on banana fruits is similar to that of snails and slugs. Thus far, it is believed that control measures will be quite expensive with population explosion.

Peel-feeding caterpillars exist in several parts of the world. For instance, *Platynota rostrana* was identified in Central America, Colombia, and Ecuador where the larvae voraciously ate up banana fruits in existing plantations (Bullock and Robert, 1961). The *Ecpanteria spp (Arctiid)* was reported in Central America showing similar defects except the black spiked sheltered epidermis which usually remained on the damaged fruit. The *Prodenia latifascia* is an artworm with a comparable eating and destructive habit to banana as die caterpillars (Harrison and Stephens, 1966). There are numerous pests and diseases, but this study covers only the major and potential ones.

Chapter Six

Banana Production Technologies

6.1 The Traditional System/Technology

The traditional system/technology refers strictly to subsistence farming. In most instances, the farm (usually small) is located at close vicinity to the owner's house. There is hardly any preference as to where cultivation should be undertaken. Further, and as Tezenas du Montcel (1987) put it, selection of the cultivated land largely depended on accessibility.

Stover and Simmonds (1987) also pointed out that the traditional cropping system is not limited to Cameroon in particular, or Africa at large, but also to some Central American and Asian countries as well.

Banana could be interplanted with cereals, rice, sorghum, chillies, sugar-canes, cassava, sweet potato, coffee, taro, maize, yams, and coconuts, if the soil is well drained (Nayar, 1962; Devos and Wilson, 1979, 1983; Remachandran, 1979; Stover and Simmonds, 1987; Stover, 1983). Furthermore, Besong (1989) pointed out that farmers in Fako prefer multiple-cropping techniques because of its better profitability margin compared to mono-cropping.

Production from this cultivation technique is locally consumed and is one of the most important sources of carbohydrates in Cameroon and other West African countries.

Tezenas du Montcel (1987) called the traditional cultivation technique "cropping without inputs", due to the non-existence of a well-defined fertilisation programme, irrigation system, drainage, and other capital investments. Planting and replanting suckers are usually obtained from a family friend, desuckered from an established garden, or purchased at little cost. In most cases, green manure, animal waste, compost, and mulching are the only forms of fertilisation available.

From a production economics standpoint, however, the animal waste or green manure, suckers, and labour expended on the piece of land are all inputs with costs, even though applied at an infinitesimal level. Further, it is difficult to quantify the amount of animal waste or green manure even man-days of labour applied. This is probably why Tezenas du Montcel (1907) labelled it "cropping without inputs".

Stover and Simmonds (1987) pointed out that in Bangladesh, India, and South China, silt is extracted underneath the canals and utilised on the banana plant. Maximum individual total cultivated area is equal to, or slightly greater than, 0.5 ha. Also, maximum yield is approximately 3 mt/ha due to the adopted intercropping system which on the other hand decreases the per ha density of the crops involved. Further, an average bunch weight ranges between 3-4 kg compared to 20-30 kg in the semi-modern and modern systems, respectively.

6.2 Semi-modern System/Technology

Approximately 85% of all the Cameroon banana firms adopt the semi-modern production technique. Stover and Simmonds (1987) pointed out that this method is used in the Windward Islands, Martinique, Guadeloupe, Surinam, Cote d'Ivoire, Somalia, and the Pacific Islands.

The technique entails a complete destruction of either a portion or the whole plantation, left fallowed, and replanted. The replanting time varies indiscriminately from plantation to plantation and from country to country. Generally, it could take place any time after three to eight years. The replanting is justified as it increases output which has been plagued either by pest and disease infestation, poor drainage systems, wind damage, poor soil structure, or inappropriate fertility programme. Moreover, replanting can be rescheduled so that harvesting occurs when there is banana scarcity, and thereafter better prices in the world market. This is referred to as the annual cropping programme (Darthenucq et al, 1987).

Stover and Simmonds (1987) indicated that farms adopting the semi-modern technology varied in sizes from 1-50 hectares and were owned by either private individuals or parastatal organisations. However, even larger plantations, i.e. from 50-1,000 hectares in size, still adopt this production technique in Cameroon, such as SPNP and OCB. These plantations have most of the characteristics of a modem or "permanent" one, such as fertility programmes, disease and pest control, road infrastructure, and so forth.

Planting methods and spacing vary from farm to farm. For instance, SPNP and Regional Centre of Banana and Plantain (CRBP) utilise the double-row planting pattern. The spacing between plants ranges from 1.5-2.0

metres whereas the "aisle" is 3.5-4.5 metres wide. The drawbacks of this pattern, as pointed out by Stover and Simmonds (1987), are the inefficient utilisation of the given "space" and "sunlight", and the difficulty of maintaining the double rows after harvesting the mother, daughter and grand-daughter plants, respectively, without sacrificing good followers. However, in the CRBP plantation, the density per ha is slightly above 2,000 plants. Maintenance of the rows should not be much of a problem since these plantations are replanted after 3-8 years.

Other planting methods include the straight-line or linear pattern, and the hexagonal patterns. The spacings between the plants vary with the musa variety.

The term, "semi-modern", in this book, simply indicates the use of modern inputs like fertilisers, nematicides, and scientific site preferences. This plantation possesses the same characteristics as the replanted small farms cited by Stover and Simmonds (1987). It should be noted that only a portion of the farm is replanted while the other is left for production, and vice versa. The idea behind this technique, according to Stover and Simmonds (1987), is simply to target favourable market prices.

However, while Martinique, Guadeloupe and other countries have successfully adopted this technique, reports from FAO, The World Banana Economy (1986) and Le Groupement d'Interet Economique Bananier (GIEB, 1988) depicted the unsuccessful implementation of the replanting technique by their Cameroonian counterpart as their peak period often occurred during unfavourable market prices.

Furthermore, and most importantly, semi-modern plantations possess absolutely no cableway networks. Consequently, the quality of fruit that arrives at the packing station is seriously hampered. In the Windward Islands and other parts of the world, the bunches are dehanded at the farm level before transporting them to the packing station. The end products are mostly for export and world market consumption (Stover and Simmonds, 1987).

Occasionally, drainage and irrigation systems are required, but due to the high cost involved, installation is impossible. Consequently, Eicher and Staatz (1985) pointed out that only 1-5 percent of total cultivable land in Africa is irrigated. Even when adopted, maintenance becomes a major problem. Consequently, the need for expatriates or highly trained technicians arises. When this is resolved, a new problem crops up, i.e. spare parts. It goes on and on and finally adaptability becomes impossible (Lowe, 1986).

Empirically, this assertion has been experienced in the irrigated banana plantations in Cameroon.

6.3 The Modern System/Technology

Finally, the modern technologically innovated plantations refer to the "permanent" cultivation technique. That is, the plantation must exist on a permanent basis. According to Stover and Simmonds (1987), "extensive technical-industrial installations must exist and are necessary for the production, marketing and sales of the fruit".

These plantations are usually owned by multinational companies (MNCs) such as the Del Monte Tropical Fresh Fruit Co., the Standard Fruit Company, and United Fruit Co. These companies are called "The Big Three". So far, only one of them, Del Monte, has established itself in Cameroon, and has indeed transferred its modern technology to it.

There were speculations that the Standard Fruit Co. (Dole) was already making a feasibility study to invest here, but this has not been publicly confirmed. Also, although Elders and Fyffes, an affiliate of United Brand, established itself in Cameroon in 1948 (Lecoq, 1972; Epale, 1985), there was no transfer of modern technology involved. Hence, modern banana technology did not exist in Cameroon until 1988.

A cableway network is one of the necessary conditions for a modern plantation. Subra (1971) and Stover and Simmonds (1987), pointed out that between 4.5-6.5 km of cableway infrastructure per 100 ha exist in some Central American modem plantations. This could be extended to even ten kilometres or more. However, the number of kilometres of cableway installation largely depend on the spaced-out distance. In the only modern plantation in Cameroon, for instance, there are approximately 178 km network with spacing of 100 meters apart.

A cableway network is a transportation system which facilitates the handling and hauling of banana bunches from the field to the packing station. The use of the cableway minimises fruit damage, such as bruises and scarring, as the bunches are separated with a spacer bar. It also reduces neck injuries, and other handling damages caused during harvesting, as the contact time is greatly reduced. The maximum distance a backer (person who carries the bunch of banana on an inflated padded tube placed on his shoulder to the cableway) carries the fruit varies between 30-50m. This depends largely on the distance between the cableways. The distance also varies from plantation to plantation, ranging between 65-100 meters apart.

In Central America and the Philippines, the fruits are transported to the packing station using either an aerial tractor, a conventional tractor, or manually (Subra, 1971; Stover and Simmonds, 1987). When an aerial or a conventional tractor is utilised, between 100-125 bunches of bananas can be hauled at a time. If manually hauled, only twenty-five bunches are allowed, on quality grounds. However, only the aerial tractor and the manual system are adopted in Cameroon.

Drainage is one of the necessary requirements of a modern banana plantation, especially in the "wet tropics" (Stover and Simmonds, 1987). It helps to leach out excessive water in the soil as it has an adverse effect on the root system, and thus the yields (Stover, 1972; Lassoudiere and Martin, 1974; Ghavami, 1976; Holder and Gumbs, 1983; Stover and Simmonds, 1987). A good drainage system comprises a principal, primary, secondary, tertiary, and surface canals, respectively. Drainage can either be by gravity, buried, or by pump. According to Stover and Simmonds (1987), the former is more frequent. However, the drainage systems in Cameroon are by gravity.

An irrigation system is an absolute necessity for a modern banana plantation for the following reasons: (a) the plantation is permanent; (b) rainfall is inconsistent throughout the year; (c) inadequate water would cause significant damage/loss due to drought; and (d) for maximum yield. In Cameroon, rainfall varies considerably. Ambe (1987) and Besong (1989) pointed out a duration of 8-9 months rainfall period (March to November) and 3-4 months dry season (November to February) in the South West Province. On the contrary, Nelson et al (1974) observed a seven months dry season in the North and a "shorter dry season" in the central part of the country.

Three types of irrigation systems exist. These are "overhead", "undertree", and "drip" (Stover and Simmonds, 1987). Appendix II depicts their advantages and disadvantages.

The overhead irrigation system is the most popular. It usually has a centralised pumping station with buried or surface pipes that feed water into the elevated sprinkler nozzles. The height varies considerably, depending on the specific type and crop to be irrigated. For instance, overhead system used in bananas are approximately 5-6 meters tall and holds between 9,500-13,000 litres of water per minute. This quantity can water approximately 150-250 ha. The undertree system pumps out approximately 13 litres per minute with a water radius of 12-18m, while the drip system supplies approximately 6 litres/hour (Stover and Simmonds, 1987; Claude Culpin, 1981; Boehlje and Eidman, 1984).

Another important requirement is a small airport and a "fungicide mixing plant". This became evident with the spread of black/yellow Sigatoka pathogen which has invaded almost all the banana and plantains around the world. Small planes, "equipped with rotary atomisers", can either be bought or hired and utilised for aerial spray, applying fungicides onto the banana plant leaves, a technique initiated in 1958 (Stover and Simmonds, 1987).

Basically, only 15% of all the Cameroon banana firms use the modern technology. The size must not be less than 50 ha but there is no maximum limit. In Cameroon, the products are sold in a protected market in France, Italy, Germany and Britain. In most instances, the EEC is the targeted market (FAO, 1986).

More than 80% of banana world trade is produced in these modern plantations located in Mexico, Jamaica, Central America, Panama, Colombia, Ecuador, Philippines, and Cameroon. Harvesting and shipment are done every week except in cases of total destruction by hurricanes or flood.

Lecoq (1972) pointed out that wind storm occurs in Cameroon at the end of the dry season and causes extensive damages. For instance, CDC lost one million bunches of bananas in 1960 and 600,000 bunches in 1964 as a result of wind storm.

6.4 Technological Revolution in Banana Production

In the past two centuries, the agricultural production in the United States went through four distinct stages. First, from 1775 to 1850, production was undertaken manually. From 1876 to 1890, that is, soon after the Civil War, animal assistance was adopted. Thirdly, from 1925 to 1950, that is, after World War I and World War II, production was mechanised. Finally, from 1950 until present, advanced technology dominated American agriculture (Knutson et al, 1983). However, the advanced technology was not introduced into the banana arena until towards the end of the 1950s and only reached its peak in the beginning of the 1970s (FAO, 1986).

Cameroon, the first wave was the substitution of Gros Michel, a variety susceptible to Panama disease caused by *fusarial wilt,* for Poyo, Grande Naine and Lacatan, a resistant "Cavendish" variety (Lecoq, 1972; Stover and Simmonds, 1987; Pans Manual, 1971; Tezenas du Montcel, 1987; Haarer, 1964). Lecoq (1972) pointed out that this disease *(Fusarial wilt)* first appeared in 1941. It was not until 1948 that it invaded the plantation in Tiko plane, hence a change of variety was inevitable. Haarer (1964) added *thai fusarial wilt* appeared early in the twentieth century, attacking all the Gros Michel cultivars all over the world. Tezenas du Montcel (1987) confirmed that all the Gros Michel plantations were affected by this disease with absolutely no effect on plantains in Cameroon, particularly, and Africa at large.

Then came the "Operation Golden Gate", when on 12th January, 1962, the United Fruit Company announced the arrival of 55,000 mt of boxed bananas from Costa Rica to San Francisco. Prior to this operation, bananas were shipped in bunches (Harrer, 1964). Further, polyethylene perforated bags and the aerial spray for Sigatoka control were adopted. Other major innovations included the cableways, irrigation, and drainage and flood system (Stover and Simmonds, 1987). These technological innovations changed banana production and productivity. Total output for the early adopters increased drastically while both marginal and average costs fell, thus increasing profitability for the early adopters (FAO, 1986).

This imported technology in the banana world has taken a quarter of a century to reach Cameroon where it is still at its initial stage. What would be the repercussions if all the Cameroon banana firms adopted these new technological innovations? This is a policy issue.

6.5 Irrigation Network

Irrigation is an important factor in banana production, even though empirical studies show that most of the established plantations in Cameroon are non-irrigated.

During a dry spell or in tropical conditions, and, based on the banana cultivations and evapotraspiration, about 1.2-1.4 times "class A pan evaporation" of water is required weekly, an equivalent of between 25-50mm compared to 0.8-1.0 times "A" pan evaporation in a subtropical weather. Further, banana is capable of absorbing between 900-1,800mm of water during the plant crop harvest, i.e. 36-40 weeks after planting (Aubert, 1968; Stover and Simmonds, 1987).

Most field damage in Cameroon, especially the "doubles" (i.e. when the pseudostem breaks in the medial point), results from drought, insufficient water and poor propping (Simmonds, 1966). This problem can partially be corrected by firmly supporting the pseudostem either with bamboos or twines. Optimum use of water is obtained when supplied twice or three times weekly with one of the irrigation systems: overhead, undertree sprinklers, or the drip, respectively. Each of these systems has advantages and drawbacks, but Stover and Simmonds (1987) rank overhead as the best followed by undertree and drip. Delayed seasonal growth is a major problem with non-irrigated plantations (Simmonds, 1966).

A 1989 field survey showed a loss of approximately 100 ha of bananas planted during the dry season (October-December) without irrigation. The plants suffered stunted and delayed growth, undersized fingers, choked bunches, smaller sized pseudostems, and very poor fruits. Studies by Robinson (1981) showed an increase from 20-30 mt/ha of bananas in Natal, South Africa, when additional water was supplied at fourteen-day intervals Further increase from 66-80% in the extra quality (the best) was recorded compared to the non-irrigated banana plantation.

Chapter Seven

The Demand for Banana

Marketing of Cameroon Banana (Demand)

The marketing of Cameroon banana depends on the following major factors represented in equation (7.1), all things being equal:

$$D_{cb} = h\ (P_d,\ P_f,\ P_g,\ Q_b,\ P_p,\ Y_f,\ E_x)\quad (7.1)$$

where:
D_{cb} = Quantity of Cameroon banana demanded by importing country
P_d = Domestic Price of Cameroon banana
P_f = Foreign Price of Cameroon banana
P_g = Prices of other fruits in the importing country
Q_b = Quality of Cameroon banana
P_p = Foreign Population and per capita consumption of banana
Y_f = Foreign Income level (disposable incomes)
E_x = Exchange rate

7.1 Domestic Price of banana (P_d)

Although Cameroon is a banana-exporting country, there is a significant domestic demand for the following reasons: (a) banana is a staple food source for some regions in the country; (b) it is cheaper than other agricultural food commodities (Almy et al, 1990; Republique du Cameroon Annuaire Statistiques, 1983). An adjustment in the domestic price will consequently affect the amount consumers will be willing to spend on banana. Economists call this sensitive reaction, price elasticity of demand. It is defined as the percentage change in the quantity demanded in response to a one percent change in price (Nicholson, 1985). Furthermore, the total expenditure on bananas will increase concomitantly with the pricing if the elasticity is greater than (negative) one, because it is an indispensable good. On the other hand, total expenditure will decline with an elasticity less than (negative) one for "luxurious goods". Finally, expenditure will remain unaltered if the price elasticity equals (negative) one.

Like all agricultural foodstuff, the price elasticity of demand for banana is inelastic. Globally, an increase in the price of an agricultural commodity has an infinitesimal effect on the quantity demanded, usually between -0.1 to -0.2, due to its extreme inelasticity (Bredahl et al, 1978; Rojko et al., 1978; Knutson et ah, 1983). However, the farm gate price elasticity in the United States ranges between -0.1 for essential foodstuffs like potatoes to -0.6 for chicken (George and King, 1971).

The inelastic domestic demand for banana is induced by culture, low income, and habit. On the other hand, world trade transactions follow the government instituted trade policies. Export trade is only possible if, and only if, the foreign price of banana is greater than domestic price, and, if the quantity for local consumption is equal or greater than an agreed minimum level (Bigman and Rentlinger, 1979a). The export price is circumscribed by world pricing, tariff, variable levies, and transportation costs (Waugh, 1944; Turaovsky, 1977; Bigman and Rentlinger, 1979). Domestic banana prices have been more or less consistent (Almy et al, 1990), even though Zwart and Meilke (1979) blame the international wheat market price inconsistency to the domestic price policies. Their model revealed that sometimes the domestic price is not correlated with the world price. It may simply be influenced by local conditions as "the balance of payments or inflation rate".

7.2 Foreign Price of Banana (P$_f$)

Cameroon banana producers have suffered from foreign price instability for a long time (FAO, 1986; GIEB, 1990). World market price instability has also been a major concern to economists for the past three decades when agricultural food commodities were plagued with consistent "stochastic fluctuations in supply and demand". An erratic price movement at any given point in time affects a certain group in the society. An increase or decrease in price, for instance, has direct repercussions on welfare gains or losses to the producers, consumers, and the government of either the importing or the exporting country (Bale and Lutz, 1981).

Price fluctuation can either be "real or policy induced" (Bigman and Rentlinger, 1979; Bale and Lutz, 1978). Equation (4.1) depicts that the production of Cameroon banana (supply) is a function of soil management, soil fertility, pest and disease control, irrigation, weather, and adapted technology. On the other hand, equation (7.1) depicts that the demand for Cameroon banana depends on domestic price, foreign price, foreign disposable income, quality, population, and so on. These uncontrollable variables in the export demand and supply equations are to be partially blamed for the erratic price movements (Bale and Lutz, 1978). Although price fluctuation can be minimised through induced stabilisation policies such as creating buffer stocks and price forecast (Just et al, 1977; Bigman and Rentlinger, 1979; Massel, 1969; Bale and Lutz, 1978), fresh banana buffer stock is impossible due to its extreme perishable nature.

7.3 Prices of other Agricultural Food Commodities (Pg)

The price of Cameroon banana in the world market is affected by the prices of other agricultural food commodities, especially close substitutes such as oranges, apples, grapes, and avocados. This price sensitivity in respect to other commodities is called "cross-price elasticity of demand" and is defined as "the proportionate change in one quantity to the proportionate change in the other price".

The cross-price elasticity can either be negative or positive, depending on whether the commodity in question is a complement or a substitute. An escalation in the price of banana, for instance, stagnating the price of oranges, may result in a rise in the quantity of oranges purchased. This is a classic example of substitute goods (Henderson and Quandt, 1980; Nicholson, 1985; Mansfield, 1985).

7.4 Quality of Export Banana (Q_b)

Quality is the most important aspect in banana export. Irrespective of the investment in the production of banana, if the product does not penetrate the market with as minimum defects as possible, a negative return to investment (ROI) is inevitable. Consumers have the right to their money's worth. Therefore, it is the responsibility of the producer to supply the best quality products as possible in order to capture the customers' trust and confidence.

Quality is not strictly limited to the fruits *per se*. It includes: the size and design of the box, the linings, the pads, the stitching, the company's logo, and even the codings. These parameters create the first positive impression to the customer. On the product side, the greenest of the fruit is the magic word as it determines the shelf-life. Further, defect severity, finger length, specified grade, packing pattern and weight are critical factors.

Even though taste is relative and quality standards vary among markets, studies on consumer behaviour show that they prefer average sized and length bananas, about 15-20 cms long. To achieve this goal, a well structured, customer oriented quality control programme, from planting to harvesting, packaging, transporting and ripening of the fruit, is an absolute necessity. Further quality evaluation before and after packaging, at the port, and after ripening, are recommended. Feedbacks obtained from these evaluations can be used to develop a strategy of quality improvement aimed at increasing the functional performance of the product, such as its *durability* (shelf-life), *reliability* (winning customers' trust, confidence, and building company's reputation and image), and *taste* (providing exactly what the customers want) (Kotler, 1980; Richert et al, 1974; Stover and Simmonds, 1987).

7.5 Foreign Population and Consumption (Pp)

The population and consumption pattern of the importing country has a major impact on the demand for Cameroon banana. Importing countries with high per capita consumption level can influence the demand for banana, but population alone is not a sufficient condition. This per capita consumption theory is only applicable to developed nations with well established food assistance programmes. The reverse is true for Less Developed Nations, LDCs (Knutson et al, 1983).

Most importing countries have multiple recipes and medicinal use for banana compared to exporting countries (Section 3.9). This explains why there is always an ever-increasing demand for the fruit. In Cameroon, and probably other LDCs, banana is only eaten green or ripe with no further use.

7.6 Income Level of the Importing Country (Y$_f$)

As mentioned earlier, population alone has an insignificant effect on demand. However, income and population play a major role in the demand for Cameroon banana. Economic theory postulates that effective demand can only be possible when backed by income. Therefore, increased income in the importing country renders the demand for banana highly effective (Knutson et al, 1983). The relationship between income and quantity of goods demanded is called "income elasticity of demand".

The income elasticity of Cameroon banana is the percentage change in the quantity of Cameroon banana purchased in respect to a one percent change in the income level of the importing country. Income elasticity can be positive for "normal" goods and negative for "inferior" goods. However, the income elasticity for agricultural commodities such as banana, apples and oranges, are usually less than one but greater than zero. This implies that a 10% increase in the real income level of the importing country will result in a less than 10% increase in the quantity of banana purchased (Nicholson, 1985; Mansfield, 1985; Knutson et al, 1983).

7.7 Exchange rate (E_x)

The exchange rate is an important but disregarded factor in agricultural marketing. Some of the earlier studies focused mainly on the macroeconomic approach such as how an increase or decrease in the US dollar, for instance, would affect the balance of payments (Kindleberger, 1968). Contemporary studies have concentrated on the effect of price and quantity with respect to a change in exchange rate (Schuh, 1974; Kost, 1976; Bredahl, 1976; Wilson and Takacs, 1976; Chambers and Just, 1978).

Kost (1976) developed a "two-country world" graphical model in which he demonstrated the effect of under and over valuation of the exchange rate to both the importing and exporting countries in a flexible exchange rate regime and how it affects the trade sector. A devaluation of the exporting country's currency resulted in an escalation in export volume, a rise in domestic price and a reduction in local consumption. On the other hand, a revaluation of the country's currency decreased export volume, increased domestic consumption, and decreased domestic price.

Schuh (1974) graphically presented, (Fig. 7.1), the effect of exchange rate appreciation in an industry supplying both domestic and world market concurrently. His analysis is more or less identical to the Cameroon banana industry, whereby both the domestic and export markets are being supplied by the same industry simultaneously. The only important exception is that his model represents a flexible exchange rate regime, whereas the latter is inflexible. In Fig. 7.1, SS and DD are the supply and demand curves. I_D is the importing country's demand for banana [in our case], assuming that the importing country cannot influence export price and that exchange rate is not volatile

Figure 7.1: Market conditions for an industry producing for both Domestic and Foreign markets

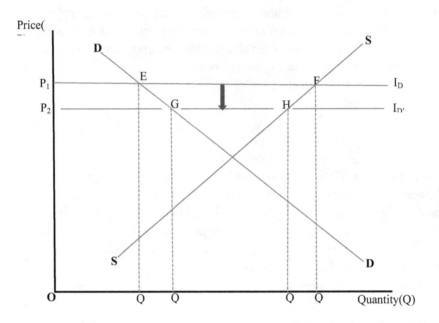

Given these conditions, the exporting country's price P_1 will be determined by the international market and the long run supply in the importing country. Q_2 represents the quantity produced whereas Q_i represents the quantity demanded locally. Q_1Q_2 represents the exported quantity at the price P, thus generating an exchange earnings of Q_1EFQ_2. Gross income generated from both the domestic and export trade would be OP_1FQ_2, whereby OP_1EQ_1, is strictly generated from the domestic market, and Q_1EFQ_2 from the export market.

An appreciation of the exporting country's currency would result in a shift from I_D to $I_{D'}$ at the domestic level, *ceteris paribus*. The repercussion to the importing country would be an increase in price in terms of the importing country's currency with a subsequent decrease in import demand.

To the exporting country, a fall in price from P, to P_2 would raise local demand to Q_3, and reduce local supply to Q_4 "as mobile resources are forced out of the industry". The quantity exported would drop to Q_3Q_4. The shaded area in Fig. 7.1 represents the export volume that has been given up for local consumption as a result of higher prices in terms of the importing country's currency or, otherwise stated, in terms of the revaluation of the exporting country's currency. Furthermore, total gross income decreases to OP_2HQ_4 with OP_2GQ_3 generated from the local transaction while Q_3GHQ_4 is generated from foreign transaction. The elasticities of demand and supply will determine to which extent the foreign earnings and gross income would decrease.

As a result of this analysis, Schuh (1974) blamed the post-World War II income problem of the United States agricultural sector on the strong dollar appreciation as it rendered the US agricultural food commodities relatively expensive to foreign consumers. He believed that the overpowered dollar contributed to the quick technological change in the US's agricultural sector, even though the advantages that came with such rapid change were not equitably distributed between the domestic and foreign consumers.

As earlier mentioned, the Cameroon banana industry produces for both the foreign and domestic markets as described by Schuh (1974), but it operates in an inflexible exchange rate system. Cameroon is a member of the CFA Franc Zone (African Financial Community). Other members include Chad, Central African Republic, Benin, Comoros, Cote d'Ivoire, Mali, Niger, Senegal, Togo, Upper Volta (Burkina Faso), Congo and Gabon (Melvin, 1985). The CFA franc is pegged to the former French franc (FF) and thus the euro, at a "par value" of 100 CFA = 1 FF=0.152449 euro, implying 1 euro = 655.957 FCFA. The Bank of Central African States (BEAC), located in Yaounde, Cameroon, and the Central Bank of West African States (BCEAO) located in Dakar, Senegal, are jointly responsible for the foreign exchange rate transactions, by ascertaining that the "actual market price" remains within the upper and/or lower one percent limit of the par value (Ingram, 1983).

According to Cunha (1984) and Nelson et al (1974), the par value 50 CFA = 1 FF was determined by the French Treasury that assured its "convertibility" in 1948. The catch is, a minimum of 65% of the CFA countries' foreign exchange reserves has to be deposited with the Bank of France. The reserve is then converted to FF, and put into each Franc Zone country's account. Withdrawal can be made at any time but the "globe account" of both BEAC and BCEAO must not be negative. Today, the exchanged is fixed at 1 euro = 655.957 FCFA.

Theoretically, BEAC and BCEAO can stockpile "unlimited overdrafts" but the French treasury charges a variable interest rate commensurate with the overdrafted amount. Furthermore, "credit balances earn interest equivalent to the going money market rate at the Bank of France". Previously, any CFA member nation with deficit balance of payment was allowed to use accumulated foreign exchange reserves. Due to the "overdraft facilities", there has been a significant decrease in the amount of foreign exchange reserves held to the disadvantage of the CFA member nations.

"In practice, the tight controls exercised over monetary policy by the joint Central Banks [BEAC and BCEAO] prevented the Zone as a whole from overrunning its own foreign reserves. Historically, these restrictions were so effective that the Zone as a whole has always been in surplus position with its operations account at the French Treasury" (Cunha, 1984).

Because the CFA is tied up with FF, it generates a false impression that it is a "hard" currency. The fact of the matter is that the CFA's worthiness is largely determined by such external forces as the French economic situation. It is not autonomous in the "International money market" and has little or no impact on the French government monetary decisions, "despite the (often ignored) provision for consultations prior to policy change and an even dimmer prospect of creating stronger national currencies".

Between 1979 and 1984, for instance, the strength of the CFA franc declined more than 55% against the US dollar. This significant devaluation destabilised the economy of the non-oil exporting Francophone African Nations by creating inflation and putting more pressure on their foreign debt transactions, especially with countries other than France. Cameroon, Congo, and Gabon were the only survivors simply because they generated extra income from their oil reserves and other resources. Empirically, a devaluation of FF against other currencies rather improves the economic relationship between France and the Francophone African Nations, as the prices of French goods are rendered more appealing. The terms of credit are positively skewed to France's advantage, rather than the rest of the world. France imports about 25% of the total exports from the Francophone African countries and exports 35% of its total exports to them. How does the devaluation and revaluation of FF affect the agricultural sector at large, and particularly the banana industry in an inflexible regime?

Since the CFA was pegged to the FF at the par value of 50 CFA = 1 FF, a devaluation of the FF means more FF are needed to purchase the same quantity of Cameroon banana, since France is the importing country and Cameroon the exporting country. Otherwise stated, it means that the CFA has been overvalued against the FF, which automatically leads to Schuh's (1974) analysis: (a) reduction of import demand; (b) increase in domestic demand; (c) fall in domestic price (P_d); and (d) increase in foreign price ($E_x P_f$) where E_x represents the exchange rate, and P_f the foreign price of Cameroon banana (Fig. 7.1).

The difference between Schuh's (1974) market and the one under study is that the Cameroon banana industry has a quantitative restriction of 60,000 mt/yr. and it operates in a preferential market. As a result, a devaluation of the importing country's currency (FF) would have the same repercussions as described mathematically and graphically (Fig. 7.2) by Chambers and Just (1978) in a simple "two-country excess demand and supply" model:

$$D_{cb} = f(P_f), D_{cb}/P_f < 0 \qquad\qquad (7.3)$$
$$S_{cb} = g(P_d), S_{cb}/P_d > 0 \qquad\qquad (7.4)$$
and
$$D_{cb} = S_{cb} = Q_{cb} \qquad\qquad (7.5)$$

where:

D_{cb} = excess demand for Cameroon banana in France
P_f = price of Cameroon banana in France
S_{cb} = excess supply of banana in Cameroon
P_d = Price of banana in Cameroon
Q_{cb} = restricted quantity of banana exported to France (60,000 mts/yr)

This model assumes no transportation costs and other trade barriers. Furthermore, it assumes that "the law of one price holds at equilibrium" as a result, and under *ceteris paribus* conditions:

61

$$P_f = E_x P_d$$

Where E_x represents the rate of exchange evaluated in terms of the units of FF per unit of CFA. Note that P_d is the domestic price while P_f is the foreign price.

In Fig. 7.2, the elasticity of the excess supply for Cameroon export bananas with respect to price is zero. That means, a devaluation of the importing country's currency (FF) will increase the foreign price (P_f), but the quantity supplied will be unchanged, due to the quantitative restriction (quota system). A revaluation of the FF, will reduce the foreign price for banana (P_f) while the quantity supplied stays constant. The domestic market will be cleared at the price P_d and quantity Q_{cb}. According to Orcutt (1950), Wilson and Takac (1976), an undervaluation or overvaluation of a currency in a fixed exchange rate regime is more stable than a short-run price fluctuation.

In Fig. 7.2, P_f and S_1 represent the foreign price and the restricted export supply while D_1 is the restricted quantity demanded. P_d, S_1 and D_2 represent the price, supply, and demand functions, respectively. A continuous devaluation of the FF, compared to other "hard" currencies in the rest of the world simultaneously escalates the prices of non-French goods on the one hand, while leaving the prices of French goods constant on the other hand. Consequently, the Francophone African countries have little or no choice but to augment import volumes from France since her prices will become relatively cheaper (Cunha, 1984).

From a welfare standpoint, both the Cameroon banana industry and the French consumers would suffer a dead-weight loss from FF devaluation. Cameroon is a net importer of most, if not all, agricultural inputs. A devaluation of the FF renders French goods more attractive in terms of prices. As a result, the reduced accrued revenue is reinvested in France in exchange for agricultural inputs. Expansion of the French banana industry and the discouragement of banana importation into France becomes possible, thus qualifying French producers as the long-run gainers of the devaluation. Furthermore, since the CFA is highly overvalued against other currencies, such as the Naira (N), consumers in these countries cannot afford to buy Cameroon bananas. This and other factors already mentioned in various sections of this study explain why the Cameroon banana firms have realised negative profit margins for the past three decades.

Figure 7.2: Market conditions for an industry producing for both Domestic and Foreign markets in a quota system

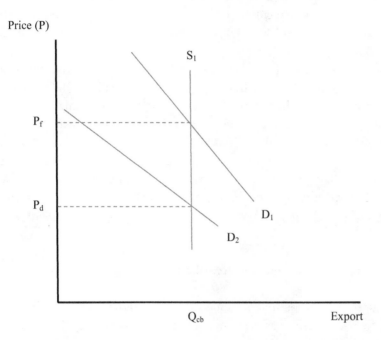

Price (P)

S_1

P_f

P_d

D_1

D_2

Q_{cb} Export

7.8 Yield Trends in the Open Market Suppliers

Banana growing modern technology was initiated in Central America. Thereafter, it was adopted by several other banana exporting industries in various countries around the world. The precise transition from "Gros Michel" to the "Cavendish" variety was not adopted in Colombia and Ecuador until half a decade later - that is, towards late 1960 (FAO, 1986).

As a result of the slow transition, the early adopters (Central America), experienced a twofold increase in yield from 20 mt to 40 mt/ha (Cochrane, 1958; Knutson et al, 1983; FAO, 1986). The Philippines recorded the same increase ten years later, whereas yields in Colombia and Ecuador were stagnated at 12 mt/ha. It was not until the 1980s that output per ha increased in both Colombia and Ecuador after the complete integration of the Cavendish banana variety and modern banana technology.

Table 7.1 shows an increase in yield/ha for both Colombia and Ecuador in 1980s, compared to 1970s. However, the data for Ecuador is inaccurate, due to the inability to determine the actual total cultivated land as a result of its "diverse production structure" (FAO, 1986). The three years averages for 1970/72 and 1980/82 in Table 7.1 show insignificant differences in yields (FAO, 1986).

Another wave of change is currently taking place in Central America where the taller Cavendish varieties such as "Giant Cavendish" and "Valery" are being replaced with the "Dwarf Cavendish" or "Grande Name". Analysts believe this new wave of change will further increase the stagnated average yields for two reasons: (a) the "Grande Maine" variety can withstand wind pressure better than other varieties; and (b) it can withstand a higher population per ha than the other varieties. However, other factors must be considered when striving for better yields, such as: soil management, drainage, irrigation, etc., (FAO, 1986; Ghavani, 1976; Lassoudiere and Martin, 1974).

7.9 Price Trends in Current and Real Terms

Amongst other tropical agricultural export commodities, such as rubber and tea, the banana price has been the most downward sloping. Estimates by the FAO (1986) show that a quarter of a century ago -i.e. from 1950 - prices in current terms fluctuated around a constant level, and real export prices fell by between 55 and 62%". Such a drastic fall in prices has a significant impact on the profitability of the banana exporting firms and/or countries.

According to the FAO (1986), the United States is the largest importer of the world's bananas. There has been no significant fluctuation in retail prices between 1947 and 1974. An average of 35.9 cents per kg of bananas was recorded during this period with a coefficient of variation of only 4.4 percent over a 25-year period (Table 7.2). As a result, the US enjoyed a consumer surplus of 18.3 cents/kg of banana purchased in real terms in 1974 compared to 1950.

Table 7.1 Selected Open Market Supplies: Trends in Yields/ha[5]

Country and Region	Three-Year Average			
	1960/62	1970/72	1980/82	1984
Central America[6]	17.4	40	41.5	-
Costa Rica	18.7	35.9	36	39
Guatemala	14	44.8	50.8	47.7
Honduras	19	42.6	50.4	40.7
Panama	16.7	41.9	36.4	48.8
South America	-	12.3	25.2	-
Colombia	-	12.7	35.2	40.6
Ecuador	12.2[7]	12.2	21.6	19.6
Asia	-	-	-	-
Philippines	-	24.58	39.2	37.5

Source: FAO (1986), The World Banana Economy.

[5] Refers to exported output per unit area only.

[6] Converted from boxes per hectare, assuming a net export weight per box of 42 lbs (19.1 kg).

[7] 1961/63 three-year average.

[8] 1973 only.

Table 7.2 Price Formation Central America to US, 1950-83

Years	Retail price (c/kg)	Import price[9]		Export price[10]	
		(c/kg) FOR	% of retail	(c/kg) FOR	% of retail
1950	35.9	16.1	44.8	10.6	29.4
1951	35.9	16.1	44.8	11.0	30.7

1952	36.2	16.3	45.1	11.1	30.8
1953	37.0	16.3	44.0	11.6	31.4
1954	37.0	16.8	45.4	11.7	31.5
1955	37.5	16.5	44.0	11.7	31.3
1956	37.0	16.8	45.4	11.8	31.8
1957	38.1	17.6	46.1	10.5	27.6
1958	38.1	16.3	42.7	10.0	26.3
1959	37.5	14.6	39.0	9.8	26.2
1960	35.1	14.3	40.8	9.8	28.0
1961	35.1	13.9	39.7	9.9	28.3
1962	35.9	13.2	39.7	11.4	28.3
1963	36.2	16.8	46.5	10.6	29.3
1964	36.4	17.0	46.7	9.9	27.3
1965	35.3	15.9	45.1	10.2	28.9
1966	34.6	15.4	44.5	10.3	29.6
1967	34.8	15.9	45.6	10.6	30.4
1968	34.2	15.3	44.8	10.1	29.5
1969	35.1	15.9	45.4	10.7	30.5
1970	35.1	16.6	47.4	10.7	30.5
1971	32.8	14.0	42.6	10.2	31.0
1972	34.8	16.2	46.5	11.1	32.0
1973	36.4	16.5	45.4	11.8	32.4
1974	40.6	18.4	45.4	12.2	30.2
1975	51.5	24.7	48.3	14.1	27.6
1976	51.8	25.7	49.6	15.4	29.7
1977	56.3	27.5	48.8	16.0	28.4
1978	62.0	28.7	46.3	17.4	28.0
1979	72.2	32.6	45.2	20.3	28.2
1980	82.9	37.9	45.7	24.3	29.3
1981	88.5	40.1	45.3	25.9	29.3
1982	78.0	37.4	47.9	25.4	32.6
1983	85.0	43.0	50.6	25.5	30.0

Source: FAO (1986), The World Banana Economy.

[9] Free on rail sales price ex-delivery shed, United States ports.

Average FOB unit values calculated from IMF export value data and country volume data. For 1950-1977 it is an average for Guatemala, Honduras and Panama. From 1978, it is an average for Guatemala, Honduras and Costa Rica.

Additionally, the report depicted that free-on-rail (FOR) prices were stable in current terms. The cost price, i.e. FOR, of an 18.1 kg box of bananas was $2.50 between 1950 and 1974. Thereafter, a 50% decline was recorded when deflated by the UN index of export unit values of manufactured goods (FAO, 1986).

There is a great dissimilarity between bananas and other agricultural products in terms of short and long run price instability. A comprehensive study of price and export revenue instability of seventeen agricultural food commodities revealed bananas to be the most stable commodity (Harris, et al, 1978; FAO, 1986) due to the high responsiveness of export supply from one market to the other in time of scarcity.

7.10 Price Formation in Marketing Channels

World banana trade takes place in two distinct arenas: open and protected markets. Trading on the free market is highly competitive, thus the production of quality products at relatively lower prices and greater output are necessary conditions for increasing a market share. Thus far, Central and South American banana producers are the main open market suppliers. They supply the United States, which is the largest banana open market, and Western Europe.

On the other hand, the protected market suppliers are given preferential treatment. Furthermore, the market is less competitive. The African, the Caribbean, and the Pacific countries (ACP) -namely, Cameroon, Cote d'Ivoire, Jamaica, and Somalia - and the French Overseas Department (DOM) - Martinique and Guadeloupe -fall under this category. Other suppliers of the Preferential Market include: Belize, Surinam, and the Windward Islands (FAO, 1986).

The FAO (1986) indicated that from 1973 to 1983, banana retail prices in the United States open market increased from 36.4 to 85 cents/kg. The free-on-rail (FOR) price also increased from 16 to 43 cents/kg while the free-on-board (FOB) price increased from 11.8 to 25.5 cents/kg. However, it is important to note the great price stability that occurred for almost 25 years until 1973, the pivotal point (FAO, 1986).

7.11 Price Trend in the Preferential Market

Similar to the open market, there has been a steady banana price increase in the protected market. The most recent increase occurred in 1988 when a kilogram of CATEGORY ONE banana was taxed at 5.32 French Francs (FF) or 266 FCFA, thus reflecting an increase of 2% (GIEB, 1988).

Table 7.3 shows the average prices for suppliers from all origins, i.e. the ACP and DOM countries participating in the French market except Martinique. Furthermore, the prices allocated for Martinique bananas are higher than other suppliers. These prices are the averages of all categories, i.e. CATEGORIES EXTRA, ONE and TWO, respectively.

Table 7.3: Price Trend in the French Market (FCFA/mt).

Year	FOB Prices		FOR Prices		Wholesale Prices (P/box)	Retail (FCFA/kg)
	Supplier from all origins	Martinique	All orgins	Martinique		
1983	3600	3695	4200	4340	5620	424
1984	3399	4133	4570	4690	6040	460.5
1985	4263	4265	4880	5040	6450	512.5
1986	4325	4455	4900	5140	6600	515.5
1987	4446	4624	4960	5220	6760	537
1988	4428	4640	5260	5400	6946	587
88/87%			0.06	0.034	2.75%	0.093

Source: GIEB (1988), Le Groupement d'interet Economique Bananier.

Table 7.4 shows price breakdown of suppliers from different origins. Martinique tops the list with the most favourable price, followed by Guadeloupe, while Cameroon is at the bottom. Both Martinique and Guadeloupe belong to the French Overseas Countries. Also, note that their 1988 prices are even greater than the 2% increase decided by the French

Interministerial decision of 21st December, 1987, setting category one price at 5.32 FF (266 FCFA/kg).

According to GIEB (1988), the large price differential between the French Overseas Departments and other countries, such as, Cameroon is simply due to the larger proportion of CATEGORY EXTRA exported by the former - hence, CATEGORY EXTRA is the best grade and sells for the better price.

7.4 Price Breakdown by origin in the French Protected Market, 1987-1988.

Origin	1987		1988		Increase 1987 to 1988
	[11]FF/kg	FCFA/kg	FF/kg	FCFA/kg	
Martinique	5.22	261	5.4	270	3.45%
Guadeloupe	5.09	254.5	5.35	267.5	5.11%
Cote d'Ivoire	4.45	222.5	5.09	254.5	14.38%
Cameroon	4.06	203	4.49	224.5	10.59%
Other suppliers	4.96	248	5.26	263	6.05%

Source: GIEB (1988), Le Groupement d'Interet Economique Bananier.

Comparative analysis, based on percentage change in price, indicates Côte d'Ivoire and Cameroon to be more favoured, as they realised 14.38% and 10.59% increases, respectively. Martinique and Guadeloupe were the least favoured as they only received 3.45% and 5.11% increase, respectively. Despite the favourable percentage change in price, Cameroon still obtained the overall lowest price in real terms compared to other producing countries (GIEB, 1988).

The reasons as depicted in FAO (1986) are twofold: (a) crop timing, and (b) poor quality. Production from the Cameroon banana industry reaches its peak point during November to January, when the French market is saturated with other fruits such as oranges, apples, grapes, and so on. Banana then faces stiff competition.

[11] Conversion factor 1FF = 50 FCFA before devaluation, January 14, 1994. After devaluation 1FF = 100 FCFA.

With the relative prices of the other fruits cheaper than banana (Table 7.5), a rational consumer's demand for banana decreases, especially at the going price. Furthermore, any consumer willing to purchase banana during this period makes sure that he/she gets his/her money's worth by requesting the best quality. It is worth mentioning that banana prices during November 1988 through January 1990 dropped 12% below the average price, i.e. from 5.05 FF/kg (252.5 FCFA) to approximately 4.50 FF/kg (225 FCFA) (GIEB, 1988).

Table 7.5 Retail Prices of Fruits in the Paris Agglomeration.

Origin	1987		1988	
	[12]FF/kg	FCFA/kg	FF/kg	FCFA/kg[13]
Pear	7.34	367	8.51	425.5
Apple	7.85	392.5	8.05	402.5
Orange	8.22	411	8.65	432.5
Banana	10.74	537	11.04	552
Grape	10.4-12.4	520-620	13.0-15.9	650-795
Lemon, Lime	11.07	553.5	10.96	548

Source: GIEB (1988), Le Groupement d'Interet Economique Bananier.

7.12 Seasonal Price Movement in the Domestic Market

According to a market study of food prices in the South-West province by the Institute of Agronomic Research (IRA), Ekona, Almy et al. (1990), banana was the cheapest food commodity in the province as the average mean price was 38 FCFA/kg from 1988-1990 compared to 88 FCFA/kg for fresh corn, 135 FCFA/kg for dry corn, 158 FCFA/kg for cassava garri, 78 FCFA/kg for plantain, 98 FCFA/kg for cocoyam, 143 FCFA/kg for white yam, and 138 FCFA/kg for the Irish potato.

Also, another study of retail prices conducted a decade ago revealed banana as the cheapest fruit in most Cameroon major cities except Bamenda (Table 7.6). The highest price (99 FCFA/kg) was recorded in Garoua, whilst the lowest (24 FCFA/kg) was recorded in Bafoussam in 1980. In 1977, 1978, and 1980, banana obtained better prices than pineapple only in Bamenda (Republique du Cameroon, Annuaire Statistiques, 1983).

Furthermore, the IRA study revealed that the most favourable banana prices were obtained in June through August, the peak of the rainy season. It is interesting, however, to note that this period of favourable prices coincided with the so called "bonne saison" when prices were driven up as a result of limited supply of banana and a mass scarcity of other food commodities. Conversely, the least favourable prices were recorded in December through February (the "mauvaise saison"), when, due to the unlimited supply of other food commodities, banana was forced to compete with other agricultural food products.

[12] FF = French Franc.
[13] 50 FCFA = 1FF before devaluation, January 14, 1994. Thereafter 100 100 FCFA = 1FF.

Table 7.6 Retail Prices of Fresh Fruits in Cameroon's Major Cities, 1977-1980 (FCFA/kg).

Item	1977	1978	1979	1980
Yaounde	48	55	52	54
Banana	249	347	361	400
Tomato	156	159	179	195
Orange	98	118	123	115
Avocado				
Bertoua	37	44	51	62
Banana	37	55	56	75
Pineapple	119	119	89	161
Avocado	131	129	179	164
Orange				
Limbe	31	35	39	61
Banana	42	55	56	75
Pineapple	57	57	76	101
Avocado	72	85	116	179
Orange	92	88	75	136
Coconut	60	57	87	103
Grapefruit				
Garoua	75	95	74	99
Banana	181	89	152	149
Pineapple	61	166	147	158
Avocado				
Bafoussam	27	25	24.8	24
Banana	103	115	131	138
Orange	33	37	42	46
Avocado	473	441	444	514
Irish				
Potato				
Bamenda	30	32	28	38
Banana	44	29	36	40
Avocado	91	84	117	125
Orange	24	30	41	31
Pineapple				

Source: Republique du Cameroun (1983), Annuaire Statistique, Ministere du Plan et de l'amenagement du Territoire.

Contrary to the domestic market price trend, the 1989 free-on-rail (FOR/kg) prices for Cameroon bananas in the French market ranged from 127 FCFA/kg to 291 FCFA/kg. The most favourable prices were obtained in March (281 FCFA/kg), April (291 FCFA/kg), and May (254 FCFA/kg) with April being the peak. Other favourable prices were recorded in September (237 FCFA/kg) and October (270 FCFA/kg). June was the worst period when the lowest price of 127 FCFA/kg was obtained.

It should be noted that the domestic and foreign markets have dissimilar characteristics, e.g. different price peaks and low periods, and different plausible reasons for the variability. In the foreign market, the peak price period falls in April, the heart of spring season, when not only is banana scarce but also its close competitors, such as grapes, oranges, lemons, and apples. The limited quantity of bananas and the unavailability of the other exotic fruits were the contributing factors for price escalation. On the supply side, the Cameroon export volume during this same period was rather downward sloping, which is contrary to the law of supply and demand. The reason, however, was because the producers were still suffering from an extreme dry spell as they had no irrigation network, except for the modern system which was still at its developmental stage.

On the contrary, the lowest price was obtained in June. The export volume was also at its lowest (Fig. 7.10). During this time, there was a boom in the exotic fruit market and banana encountered fierce competition. Furthermore, this was the beginning of summer, when temperatures rose from 80°F to 100°F (27°C to 38°C). This abrupt change in temperature, coupled with other factors such as the availability of other fruits and the consumers' burning desire to change taste, all contributed to pulling down the price of bananas in the world market.

The second price peak (local or relative maximum), was in November, when the weather started getting cold again (Autumn) and the other fruits began to disappear. On the supply side, export volume rose sharply due to the transitional change from rainy season to dry season.

Generally, banana growth is highly restricted during extreme rainfall and inadequate sunshine periods. Consequently, June to September is usually unfavourable to the Cameroon banana growers. However, the situation is overturned from October, the beginning of the dry season, as the hours of sunshine per day increased. The semi-modern systems were most affected during this period of absolute minimum production and prices in terms of profitability, that is, from June to September.

From an economic standpoint, profitability could still be maximised if and only if export volume was increased during the price and output slump period by adopting a strategic farm management technique.

This technique in quite complex and requires the following minimum historical information for successful implementation:

(a) Rainfall data of the plantation area for at least 5-20 years;

(b) 5 -10 years sunshine data;

(c) Improved drainage system in the required areas in the plantation;

(d) Plant population adjustment based on the soil type, texture and class;

(e) Reinforcement of the pruning operation; *and*

(f) Maximisation of fruit age.

If this information is available, a competent and well qualified farm manager can successfully adjust the downward production trend to the company's advantage.

7.13 Marketing Structure and Costs

Harrison (1987) defined marketing structure as "the physical infrastructure that supports marketing institutions, the government policies and programmes supporting or hindering marketing institutions, and the mix of marketing institutions (types of marketing businesses) providing the marketing services". Kotler (1980) took the following parameters into consideration: (a) the size, strength and position of the firm in the market; (b) the firm's liquidity, goals and future plans; (c) the purchasing power and attitude of the consumers; and (d) the economy at large.

Approximately 80% of banana consumers in the Douala area come from all classes of the Bamileke ethnic group, who, because of lack of habitable land in their original territory (Western province), migrated to the Coast in the 1930s. By 1970, the population density had risen to 143 inhabitants/km^2 with a 3% birth rate (Nelson et al, 1974). Furthermore, 40% of the population practise polygamy with an average of three wives (Republique du Cameroun, Annuaires Statistiques, 1983). Their dynamism in business and other activities and their natural money management skills have earned them the title: "the Jews or the Ibos of Cameroon" (Mbuagbaw et al., 1987). They have dominated all the significant business activities in Douala and other major cities of the country as cab-drivers, educators, doctors, and civil servants (ministerial positions). According to Nelson et al (1974), 70% and 30% of professional and civil service jobs were occupied by this ethnic groups in Douala alone in the 1970s. Furthermore, 60%, 80%, 40%, and 12% were traders, artisans, labourers, and domestic servants, respectively. It is likely that these percentages may have significantly changed in the 1990s.

Two techniques are adopted in the calculation of marketing costs. The contribution-margin technique, in which only the direct expenses are incorporated; and the full-cost technique, which simultaneously incorporates the direct and indirect expenses (Stanton and Burkisk, 1978). Only the full-cost technique was utilised to net the returns of the wholesaler/truck-load of banana sales.

Part Three

Research Methodology

Chapter Eight

Methodology

Introduction

This book comprises two major components: production and marketing. First, it focused on banana production under three different technologies: modern, semi-modern, and traditional. Then, the marketing component involved foreign and domestic price analysis, trends in banana export, distribution channels, market structure, and marketing costs assessment.

8.1 Sampling Plan

The various objectives for this study called for more than one sample of respondents. A sample of markets was needed to gather information about aspects of the marketing objectives. Samples of models of production were taken to gather information on the production and farm gate prices, respectively.

8.2 Production Respondents

A total of 53 producers were selected and interviewed from the three classes of banana-producing firms adopting either modern, semi-modem, or traditional technologies, respectively.

The only firm adopting modern technology, CDC/Del Monte (DMC), was purposively selected. This firm is located in the Tiko plane of the South West Province (Appendix III).

Two out of the three firms utilising semi-modern technology were randomly chosen. The number was reduced to three because the Societe des Plantations de Nyombe-Penja (SPNP) and Plantation des Haute Penja (PHP) were both owned by the same French-based parent company, Fruitiere, and thus adopted the same technology. Therefore, separating them would be double counting (Appendix III).

It should be noted here that two of the three firms - Organisation Camerounaise de la Banane (OCB), and SPNP/PHP - were based in the

Littoral Province, while the Cameroon Development Corporation (CDC) Ekona was located in the Southwest Province.

One of the two firms in the Littoral Province was selected, based on the willingness to cooperate and/or participate. Conversely, CDC Ekona was purposively selected since it was the only firm adopting semi-modern technology in the Southwest Province.

Fifty smallholders adopting traditional technology were randomly selected (twenty-five from the Littoral and twenty-five from the Southwest Provinces respectively).

8.3 Marketing Respondents

A total of 53 intermediaries were selected and interviewed. Out of these, three were wholesalers from the modern production system while fifty were retailers from the three different systems in question. The three wholesalers were purposively selected while the retailers, five from each market (see Section 9.3) were randomly selected and interviewed.

Chapter Nine

Data Collection

Data for the three different production systems were collected from both secondary and primary sources. Secondary source data were from records of various ministries and international organisations. The Ministries of Agriculture, Trade and Commerce, Finance, and Scientific Education, the World Bank, Food and Agricultural Organisation of the United Nations, United States Department of Agriculture, and Institute of Agronomic Research typify some of the secondary data sources. On the other hand, primary data were obtained from production and marketing questionnaires.

9.1 Production

Questionnaires were sent out to the production managers and/or owners of the fifty-three selected firms representing the three production technologies under study. Relevant data on banana production and farm gate prices were also collected from the questionnaires.

On banana production, data was collected on input application and costs, such as fertilisers, chemical weed control, labour cost, output per ha, price per box of bananas, price per kilogram, cost of aerial spray, cost of pest and disease control, cost of irrigation and other fixed inputs. Secondary data were also used to complement information that could not be obtained from primary sources. These data were used to compare banana production under modern, semi-modern, and traditional technologies and compare foreign and domestic prices of bananas.

9.2 Farm Gate Price Survey

The economic and financial values of the 'rejects' reflect a better estimate of the farm gate price, assuming a perfectly competitive market. Otherwise, an adjustment of the opportunity cost of using the rejects is needed to determine the farm gate price (Gittinger, 1984).

'Rejects' simply refer to bananas that for one reason or the other did not meet export/world market standards. For instance, if the length of the fingers was slightly shorter than the minimum specification (19 cm for GRADE EXTRA and 17 cm for GRADE ONE if measured from the outer-whorl), or if the size was smaller than the minimum required calibration (diameter), i.e. 34 mm for GRADE EXTRA and 32 mm for GRADE ONE, if measured in the middle of the fingers, the cluster or hand would be rejected even though the general quality was not hampered. Furthermore, the quality requirement varied from market to market. Consequently, a rejected quality in one market could well be accepted in another.

The farm gate price is usually the most acceptable value for a "home-consumed production", since it is argued that the real worth of the crop is over-valued when the market price is used as an indicator, especially when only a negligible quantity of the agricultural food commodity is actually sold locally. As a result, the farm gate price portion of the questionnaire included questions such as: how many kilograms of rejected bananas were obtained each harvesting/ packaging day; how many harvesting/packaging days occurred weekly, monthly, or yearly; how many kilograms were sold each week, month or year; what was the price per kg in FCFA; how many kgs were re-utilised as production input; and what was the approximate yearly income generated from the sale of rejects.

Information on the farm gate price was also obtained from secondary sources, such as accounting records and receipts from the daily, weekly, and monthly sales of rejects to middlemen from the organisations under study. Additional information was obtained from personal interviews of the intermediaries. This study incorporated the farm gate price in the cash flow and/or profitability analysis, sensitivity analysis and marketing costs analysis as they reflected the actual revenue generated by the producer firms from the sales of their rejects.

9.3 Marketing Survey

Ten domestic major markets were selected and surveyed. Three out of the ten (Tiko, Mutengene and Muea) in the Fako division were purposively selected because of their sizes and nearness to the main plantations. Three in the Mungo division (Bouba III, Nyombe, and Penja) were randomly selected. Finally, four major ones (Deido, Bonaberi, Nkololo, and Cite-Sic) were randomly selected in Douala where most of the bananas from both semi-modern and modern plantations were being sold. Furthermore, a randomised selection of 250 consumers, 25 from each of the ten local markets, was carried out.

The data for the foreign marketing survey were obtained partially from secondary sources such as libraries, some government ministries (i.e. finance, transport, commerce and industry), agriculture and trade associations. More information was obtained from international organisations involved directly or indirectly in banana export operations and/or research. The World Bank, the Food and Agricultural Organisation (FAO), and the Institute of Fruits and Agronomic Research (IRFA.) typify such organisations. Also, primary data were extracted from the production questionnaire and personal

This survey comprised data on weekly, monthly and yearly export volumes of bananas for each of the selected firms under the given technologies, and were used to analyse the seasonal and annual trends of banana export. It further provided information on distribution channels, market structure, and the costs incurred in marketing, e.g. transportation of fruit from the farm to the ports, harvesting and packaging costs, stevedoring and other charges. The above data were used to analyse the seasonal and annual trends in volume of banana export; describe the distribution channels and marketing structure of bananas; and, assess marketing costs in banana trade.

Chapter Ten

Data Analysis

For banana production (supply model) and marketing (demand model), descriptive statistics such as means, percentages, and graphs were used to analyse the data obtained from the production and marketing questionnaires. Financial analysis such as capital budgeting, income and expense statements were also used to determine profitability levels of the various banana production systems, and the various intermediaries who facilitated the chain of distribution, from the farm gate to the final consumer. A comparative costs and benefits, and sensitivity analysis of the three production systems were computed.

10.1 Banana Production (supply) and Marketing (demand)

The banana production (supply model) was formulated in Chapter Four as:

$$S_{cb} = g(S_{mg}, F_{tp}, P_{dc}, I_{rr},...,Z_n) \qquad (10.1)$$

where:
S_{cb} = the export supply of Cameroon bananas (Mt/ha exported);
S_{mg} = Soil management (FCFA/ha);
F_{tp} = Fertility (FCFA/ha);
P_{dc} = Pest and disease control (FCFA/ha);
$I_{rr.}$ = Irrigation network (FCFA/ha);
Z_n = other production inputs and factors such as weather, adopted technology and management (FCFA/ha).

was implicitly incorporated in the empirical results of Chapters Eleven to Seventeen.

For instance, in the comparative cost-returns analysis, it was assumed that in year one (Y_1), 8, 17 and 33 mt of banana/ha were produced in the traditional, semi-modern and modern technological systems, respectively. Since every agricultural food commodity requires several inputs, and in order to avoid repetition, it was implicitly assumed that equation 10.1 was adopted either partially or wholly in the attainment of the given yields.

By definition, the supply curve of Cameroon bananas explains the relationships between the available output disposed at various market prices. The supply function is seldom derived in mathematical economics because in a competitive market and under profit-maximisation condition, the firm's supply curve is represented by the section of the marginal cost curve beyond the average variable cost curve, where price is equal to marginal cost (Teh-wei Hu, 1980).

On the other hand, a similar implicit assumption was in the marketing or demand model formulated in Chapter Seven as:

$$D_{cb} = h(P_d, P_f, P_g, Q_b, P_p, Y_f, E_x)$$ (10.2)

where:
D_{cb} = Quantity of Cameroon banana demanded by importing country;
P_d = Domestic price of banana;
P_f = Foreign price of Cameroon banana;
P_g = Prices of other fruits in the importing country;
Q_b = Quality of Cameroon banana;
P_p = Foreign population and per capita consumption of bananas;
Y_f = Foreign income level (disposable incomes)
E_x = Exchange rate.

For instance, the quantity of Cameroon bananas exported at any given time faces fierce competition, not only from banana from other parts of the world but also from other fruits such as grapes, apples, and citrus, in terms of prices (P_g). Market clearance depends on the income (Y_f) and/or the purchasing power of the importing country's residents (P_p). Furthermore, any business transaction or cash payment that involves two or more countries utilising different currencies will generate the demand for foreign exchange. The Exchange rate is "simply the price of one nation's money in terms of other currencies (Houck, 1986; Ingram, 1983; Kenen, 1985).

10.2 Capital Budgeting Analysis

A net cash inflow of bananas produced per ha under the three given technologies was done over a 10-year period. The three different technologies, i.e. modern, semi-modern, and traditional, were considered as independent projects. Consequently, the acceptance of one, in terms of profit margin, did not eliminate the others since the projects were not mutually exclusive. Estimating cash flows is one of the most important and difficult steps in analysing capital budgeting. It requires many variables and critical assumptions (Brigham, 1982). The data obtained from the production questionnaire and other secondary sources were utilised in the net cash inflow analysis. Two of the most sophisticated capital budgeting techniques, the Net Present Value (NPV), and the Internal Rate of Return (IRR), were adopted using the following formulas:

$$NPV = \left[\frac{CF_1}{(1+K)^1} + \frac{CF_2}{(1+K)^2} + ... + \frac{CF_n}{(1+K)^n} \right]$$

$$= \sum_{t=1}^{n} \frac{CF_t}{(1+K)^t} - C$$

$$= CF_1(PVIF_{k,1}) + CF_2(PVIF_{k,2}) + ... + CF_n(PVIF_{k,n}) - C$$

where:
CF_1, CF_2,..., CF_n = annual net cash inflows;
K = the approximate discount rate depending on the level of risk involved in the project and the economy's interest rate;
C = the initial cost of the project;
n = expected life span;
$PVIF_{k,n}$ = present-value interest factors for one franc CFA discounted at k percent for n periods.

The Net Present Value (NPV), according to Brigham (1982) and Gitman (1982), is the "present value of the expected net cash flows of an investment discounted at an appropriate percentage rate, less the initial cost outlay of the project".

In a layman's terminology, and as Gitman (1982) puts it, the NPV is simply the present value of net cash inflows minus the initial investment. A proper comparison of an investment portfolio will be impossible if both the cash inflows and outflows are not stated in its current value. Furthermore, the "accept-reject" decision criterion is based on whether the NPV is greater or equal to zero, i.e. if NPV $\geq$ 0, the project is accepted. On the other hand, if the NPV $<$ 0, the project is rejected.

Brigham (1982) and Oilman (1982) define IRR as "the discount rate that equates the present value of the expected future cash flows, or receipts to the initial cost of the project". Stated otherwise, "it is the discount rate that equates the NPV of an investment opportunity with zero (since the present value of cash inflows equals the initial investment)". The following formula was used for computation:

$$NPV = \left[\frac{CF_1}{(1+r)^1} + \frac{CF_2}{(1+r)^2} + ... + \frac{CF_n}{(1+r)^n} - C \right] = 0$$

$$= \sum_{t=1}^{n} \frac{CF_t}{(1+r)^t} - C = 0$$

$$= CF_1(PVIF_{r,1}) + CF_2(PVIF_{r,2}) + ... + CF_n(PVIF_{r,n}) - C = 0$$

where:

CF_1, CF_2,..., CF_n = annual net cash inflows;

C = the initial investment;

r = internal rate of return;

$PVIF_{r,n}$ = present-value interest factors for one franc CFA discounted at r rate of return for n periods.

Besides the rate of return, r, all other variables such as the annual cash inflows CF_1, CF_2,..., CF_n, and the initial investment, C, were determined from the production questionnaire data and secondary sources. Therefore, the adopted IRR equation had one unknown variable, r, which had to be determined.

The "accept-reject" decision criterion was based on whether or not the internal rate of return was greater than or equal to the cost of capital. The project was accepted if the IRR $\geq$ cost of capital, otherwise it was rejected. Several techniques can be used to calculate the IRR. The trial and error, the graphic, and the constant cash inflows methods typify some of the available techniques. However, a computer was used to determine the value of r in this study.

10.3 Comparative Analysis of Costs and Returns

A comparative analysis of costs and returns per ha of bananas produced under traditional, semi-modern, and modern technologies, respectively, was done. The data utilised were extracted from the cash flow or capital budgeting analysis in Tables 15.1, 15.2, and 15.3, respectively.

Furthermore, a tabulated input cost analysis for the three different technologies was computed. It facilitated the determination of various input costs as a percentage of total cost (TC) of production.

10.4 Sensitivity Analysis

A sensitivity analysis for the three production systems was calculated. This analysis indicated exactly to what extent the net present value (NPV), the internal rate of return (IRR), and the pay back periods (PBP) fluctuated in response to a change in any input variable, holding other things constant. The cash flow or capital budgeting analysis for the three different technologies illustrated in Tables 15.1, 15.2, and 15.3, respectively, and discussed in Chapter Fifteen, were considered the "base case" situation. The base case input variables chosen were sales volumes, sales price, operating costs, labour cost, packaging material, aerial spray, and freight. These input variables were changed by the following percentages above and below the base case value: ±5%, ±10%, ±15%, and ±20%. The base case value was represented by zero percent. The reactions of the NPV, IRR, and PBP, in response to the percentage changes in the input variable, were recorded.

Furthermore, a graphical sensitivity analysis on sales volume for the three technologies, modern, semi-modern, and traditional, was displayed. The slopes of the curves showed how unstable the NPV, IRR, and PBP were in response to a percentage change in sales volume. The steepness of the sensitivity lines indicated the degree of risk associated with each of the production systems. This is an important tool for modern management. It is worth mentioning that an insignificant change in sales volume or prices can easily change profitability or the NPV, IRR, or PBP from positive to negative and conversely.

Conceptually, and according to Sassone and Schaffer (1978) and Brighara (1982), there are three types of sensitivity analysis: subjective estimates, selective sensitivity analysis, and general sensitivity analysis. The subjective estimate is the easiest and fastest approach. It requires "experience, intuition, and gut feeling". The major disadvantage of the subjective estimate is the manager's inability to back up his/her decision if questioned.

In the selective sensitivity analysis the analyst selects input variables that might be subjected to errors or which might have a significant effect on the NPV, IRR, and PBP, if altered. Likely "high and low" or "best and worst" values, in percentage terms, are assigned to the input variables. New NPV, IRR, and PBP values are then computed with each of the assigned values. The superiority of this approach is based on its objectivity, computational ease, and available software, but critics believe unskilled managers may be unable to make a sound decision if confronted with several choices.

The general sensitivity analysis requires a computer simulation which assigns a probability distribution to each of the main parameters in the analysis and monitors the behaviour of the NPV, IRR, and PBP. The advantage of this approach is that, if correctly computed, even an unskilled manager can determine profit margin of each of the three production systems at a glance. It also provides complete and reliable information that is easy to interpret. However, it is very rigorous, time-consuming, and requires special computer software. Theorists rank the general sensitivity analysis as the best, if and only if the tune constraint could be limited, and the software designed. The selective sensitivity analysis is recommended next, whilst the subjective estimate is the least favoured. However, the selective sensitivity approach was adopted for this study because of its objectivity, computational ease, and available computer software.

10.5 Export Volume and Price Analysis

Descriptive statistics such as graphs, tables, percentages, means and histograms were used to derive trends in export volume and price.

A ten-year trend analysis for Cameroon bananas exported to France and Britain was drawn up. Furthermore, an analysis of the percentage change in export volume, comparing the semi-modem and the modem era was tabulated. A graphical presentation of fourteen years net export to the world market was made to reveal the best and the worst years in the Cameroon banana economy. A monthly graphical analysis of the four major French market suppliers, Cote d'Ivoire, Cameroon, Guadeloupe and Martinique, was drawn up to determine the most favourable trade nation.

Also, a monthly price trend for the four major suppliers of the French market was drawn up to determine the most favourable nation in terms of price (FOR) per kg of bananas and, when the best and worst prices were obtained. Furthermore, a monthly comparison of export volume and price for the four major producers was made to show the most favourable country in terms of trade practices, i.e. maximising output when prices were at their best by optimising overall export volume.

10.6 Domestic Marketing Analysis

To analyse, describe, and summarise the data from the marketing survey, an income and expense statement per tractor-load of banana sales by the wholesalers of bananas from the modern production system, using the full-cost approach, was estimated. The income and expense statements per tractor-load of banana sales by the retailers of products of both the modern and semi-modern production systems, adopting the full-cost approach, were also calculated, and net profit variability shown. The full-cost approach was utilised because it included all the direct and indirect expenses absorbed by the intermediaries, which facilitated the determination of their net profitability in the reject banana business.

Generally, it was a cumbersome task to derive accurate marketing costs incurred by intermediaries since monitoring their routine daily activities was practically impossible. The determination of cost classification into direct and indirect was also a major problem. Costs that were directly linked to reject banana sales were classified as direct or separable costs. On the other hand, those with no direct links to the reject bananas were categorised as indirect or common costs. Frequently, the entire or a segment of the distribution costs fell in the latter category. Finally, the determination of a marketing cost was an expensive and time-consuming operation.

Additionally, descriptive statistics such as percentages and means were utilised to analyse the data from the consumer marketing survey.

91

10.7 Limitations of the Study

The following major drawbacks were encountered during this study: (a) data limitation; (b) insufficient literature on Cameroon banana production and marketing; (c) financial constraints; (d) time constraint; and (e) lack of cooperation from corporate producers, traditional farmers, and intermediaries.

Data collection was a handicap. Most of the corporate banana producers in the semi-modern system were reluctant to release professional information and respond to questionnaires. Those who were willing to cooperate had to obtain permission from higher management, which in most cases was not easily granted. As a result, each piece of information required several contacts and trips. Collecting data and/or information from various ministries was indeed a painstaking task. Even when permission was granted from top management, the officer in charge was unwilling to go through huge stacks of old and dirty files without compensation and for fear of staining his/her clothing. Whenever a file was successfully retrieved, the information and/or data was incomplete as there were always several missing papers/pages. Consequently, a set of complete information/data required several sources, contacts, time, and finance.

Literature and/or certain data on Cameroon bananas and marketing were scarce. The available studies were done by either foreign individuals or international organisations. Ordering them from abroad was expensive and required a lot of patience and time. On several occasions the books or documents were missing at the post office, thus slowing down the pace of the work.

Financial constraint was a major setback. No financial aid was available and/or granted for this research. All costs incurred (i.e. computer, stationery, field trips, travel expenses, etc.) were absorbed by the authors.

Finally, obtaining accurate information from the intermediaries (wholesalers and retailers) was a problem. The reasons were twofold: (a) they were afraid that the interviewer might be a member of management staff in either of the production systems and might raise prices if he or she found out that the profit margin was substantial; and/or (b) they suspected the interviewer to be a prospective competitor, so they tried to discourage possible entry into the industry by sometimes releasing false and negative information. As a result, any piece of information required several verifications and authentication. Nevertheless, the steps taken to overcome limitations made it possible to obtain usable data for this book.

10.8 Suggestions for Further Research

Most of the limitations enumerated in Section 10.7 are subjects for further research. No effective research can be done without available data, literature, and finance. An information and documentation system for bananas, plantains, and other agricultural food commodities is a necessity and should be one of the government's priorities, especially the Ministry of Scientific Research. However, to meet the immediate needs of the Cameroon banana firms, future research should be focused on the following areas:

(a) Soil management;
(b) Agricultural engineering with emphasis on water requirements, irrigation and drainage;
(c) Plant nutrition and fertilisation;
(d) Disease, pest and weed control;
(e) Farm management and production systems from an agro-economic approach; *and*
(f) Storage, ripening and banana processing.

Part Four

Empirical Issues In
Banana Production and Marketing

Chapter Eleven

Traditional Technology

11.1 Cultural Practices in Traditional Production Technology

As earlier mentioned in Section 6.1, the traditional production system is subsistent farming. The farm is usually located near the farmer's house. The size varied between 0.25 and 1.0 ha. Generally speaking, there is hardly a traditional banana monocultural cultivation system in Cameroon. Out of the fifty respondents interviewed during the study (twenty-five from the Southwest and twenty-five from the Littoral provinces, respectively), none of them adopted mono-banana farms. On the contrary, multiple-cropping was very common due to the "greater production per unit area compared to monoculture" (Lowe, 1986; Besong, 1989). Banana was intercropped with plantain, cassava, yam, groundnuts, corn, cocoyam, coffee, cocoa, etc., in this study.

11.2 Land Preparation

The choice of land depended on availability and was non-scientific, i.e. no soil test or analysis was done. Land preparation immediately accompanied land acquisition. The need for additional labour depended on the size and the number of large trees in the acquired farm. Unsophisticated equipment, such as machetes, diggers, hoes, and files were the farmer's capital goods. Most of the time, the farmer did all the tree felling, clearing, mounding, and burning required. An average of eight man days per ha were allocated to land preparation. The wife and children might assist in hoeing, weeding and ridging.

11.3 Planting, Planting Material, and Plant Population

Planting begins in February for other food crops such as cassava, maise, yams, groundnuts, etc. (Besong, 1989), and mid-March/early April (beginning of rainy season) for bananas and plantains. Due to the multi-culture practices, banana and plantains were averagely spaced out at 5 metres by 5 metres. Based on the spacing dimension, one would expect at least 400 mats/ha but, empirically, 339 mats/ha was the average. It was further noticed that traditional farmers preferred plantains to bananas as an average of 5.6 percent of the total mat/ha of the farms visited was bananas, while the rest was allocated to plantain. The planting pattern was haphazard.

Furthermore, traditional farmers preferred small sword suckers to the maiden and bullhead suckers, simply because the former was easier to transport from one location to die other. After slight peeling, planting was done in a hand-dug hole without cutting the pseudostem. The size of the hole recommended was 60 cm by 60 cm by 60 cm (Tezenas du Montcel, 1987), and the top soil was first put back in the hole, then the sucker, and finally the subsoil. However, farmers using the traditional production system did not adopt all of these recommendations.

11.4 Pruning, Replanting, and Plant Population Count

Stover and Simmonds (1987) defined pruning as "the technique of selecting the most vigorous sucker in the best location with respect to adjacent mats, and eliminating the undesirable suckers". The traditional farmer did neither pruning, replanting, nor plant population count. Occasionally, at the request of a family friend, desuckering for replanting or planting in another farm may occur. Pruning is one of the most vital operations in banana production, even though it is ignored in the traditional cultivation system.

11.5 Propping

Propping is simply the provision of adequate support to fruiting plants. Three types of bamboos or poles propping were common in the traditional system. First was the mono-forked prop placed against the "floral stalk" and firmly buried in the same direction as the emerging bunch. Then, the "vertical prop" which was tied against the floral stalk and buried parallel to the emerging bunch. Finally, the scissors propping, in which both ends of the bamboos were tied against the emerging bunch and the bamboos/poles placed at a 45^0 angle forming a "tripod".

11.6 Deleafing

Deleafing is the elimination of broken, disease-infected leaves, and leaves touching the emerging bunch. The operation was seldom done in the traditional cultivation system. Occasionally, the farmer or his wife needed the leaves for cooking. Instead the good leaves needed for plant growth were cut off. Tezenas de Montcel (1987) pointed out that hanging leaves protected the plant from "sun rays" during dry spell. Conversely, they provided hiding facilities for adult weevils in the base of the pseudostem.

11.7 Bagging, Debudding, and Dehanding

Bagging is the covering of the bunch with a polyethylene perforated bag in the early stage of shooting. Debudding is the elimination of the male bud about 10 cm from the index finger to avoid rotting. Dehanding is the field removal of the false hand, plus one or two true hands, or as recommended. These three operations are done simultaneously. The advantages include: yield improvement and fruit quality, increased finger length, fruit maturity time, and protection from damage by insects, bats, and birds. These cultural practices were not identified in the traditional production system.

11.8 Harvesting

Harvesting was done by the farmer alone when necessary and only a machete was needed. After removing the prop (if any), a "V" mark was inserted on the pseudostem. The bunch slowly descended without touching the ground. The farmer then cut off the bunch. Some farmers completely eliminated the pseudostem while some cut it halfway. The bunch was sold either at the local market in front of his door or was home consumed. Sometimes it was ripened before being sold.

11.9 Weed Control, Nematicide, and Fertiliser Application

Chemical weed control was seldom, but manual weeding was common - two to three times a year - or when necessary. There were neither nematicide nor fertiliser applications.

Chapter Twelve

Semi-Modern Technology

12.1 Cultural Practices in the Semi-modern Technology

There were some similarities between the semi-modern production system and the modern system. The dissimilarities were directly associated with the natural physical features, such as the topography and soil structure. Section 6.2 covered the distinguishable characteristics in detail. This production system was strictly monocropping.

12.2 Land Preparation

Despite the uneven topographical features in a semi-modern production system, land preparation was done mechanically and manually. Since replanting occurred between three and eight years, this production system was classified under "rehabilitated plantation". Thus, land preparation followed the following sequence under normal circumstances:

(a) All the old drainage was restored (if available);

(b) There was the re-establishment of the irrigation network (if available). It should be noted that less than 25% of all the farms adopting the semi-modern production system utilise irrigation in Cameroon);

(c) A nematode/borer survey was undertaken;

(d) The unwanted bananas were destroyed with machetes and/or bulldozers;

(e) The cleared area was deserted for a minimum of five months to eliminate nematodes and borers;

(f) Herbicide was applied to eradicate unwanted weeds; *and*

(g) The planting operation began (Stover and Simmonds, 1987).

12.3 Planting Material

Seed propagation was obtained from three sources: (a) a seedbed nursery; (b) extraction from existing plantations; and (c) the "in vitro" system.

Generally, 10 suckers/plant can be produced from a splendidly sustained seedbed. A population of 1,700-1,900 per ha, which can produce enough seeds for approximately eight ha, is recommended (Stover and Simmonds, 1987). The "leaf stripping" or "bud exposing" technique recommended by Barker (1959) is used to develop about 20 suckers per mat in less than a year in a seedbed. Another method is the cutting back of the mother plant after the emergence of the peepers to stimulate growth. It is important to inhibit the regrowth of the mother plant by killing the heart with a machete.

The extraction from an existing plantation technique involved the intentional development of "doubles", i.e. the mother and two daughters instead of one daughter for a production unit. Consequently, a daughter was extracted when it "reached a <u>minimum</u> of 15 cm diameter and 15 cm above the soil (Stover and Simmonds, 1987). Extraction was delayed until the mother plant was fruiting or harvested, to avoid destruction of the root system.

The in-vitro micropropagation, commonly called meristems, is the latest technique in which about 4,000 plants/day/person, 5-10 cm tall, can be produced in laboratories. These plants are further nurtured in a nursery for approximately three months before being transplanted to the field. It is believed that the meristem culture is free from soil-borne diseases (Berg and Bustamante, 1974; Stover and Simmonds, 1987). However, no semi-modern firm has attempted this technique in Cameroon.

12.4 Planting

Planting methods and distances were discussed in Section 3.3.2. More than 80% of the semi-modern production system adopted the mechanical planting technique. Trenches of 60 cm deep were excavated with tractor drawn ploughs. A string with a stake attached on each end was tensioned. The recommended planting distance was labelled on the string (Section 6.2). The planters simply lay the plant on the labelled/marked position prior to the planting day. Earthing up was done the following day with the plant in the upright position using a hoe or shovel. The top portion of all the plants were cut off. The process continued until planting was complete.

Three types of seeds were usually available. The small sword suckers, the maiden suckers, and the bullheads. Most farms using the semi-modem

102

production system in Cameroon preferred the small sword suckers as planting materials.

12.5 Pruning, Replanting, and Plant Population Count

Pruning was essential to maintain the double row planting system. The opportunity cost was the good seeds forgone in order to preserve the rows. Generally, pruning minimises scuffling for the available soil nutrients, water, and sunlight among the mother plants and the followers. Group pruning was common. A broad machete and a sharpener were utilised. Replanting and plant population count/control were seldom done in the semi-modern production system.

12.6 Propping, Deleafing, Bagging, Debudding and Dehanding

Initially, the bamboo/pole propping technique was the common practice in the semi-modem production system. Adoption of the twine method was gradually replacing the former. Deleafing frequency was irregular except in the Sigatoka-infested areas. Approximately 50% of the semi-modern plantations have adopted the bagging, debudding, and dehanding operations. The dehanding varied from false plus one to false plus three.

12.7 Irrigation

Even though water is an essential factor in banana production, less than 25% of the semi-modern plantations in Cameroon were irrigated. Overhead canons were installed all over the irrigated farms. These canons were equipped to supply about 120 mm of water every month.

12.8 Weed control

Chemical and manual weed controls were practised. This is very critical in banana production as weeds compete with bananas for growth factors such as light, nutrients, water, oxygen, and carbon dioxide (Akobundu, 1987; Aldrich, 1984; Zimdahl, 1980). Knapsack sprayers were used after a mixture of the recommended chemical and dosage. The chemicals commonly used were gramozone, basta, armada, gramuron and round-up.

Conversely, handslashing was used on "overgrown annual weeds". The repercussions of this technique in banana cultivation were threefold: (a) if the handslashed area is not backed up with chemical control, an accerated regrowth occurs; (b) the slashed weed continues to compete for light, water and soil nutrient since the root system is still intact; and (c) an increased manpower is required for such a tedious operation, thus escalating production costs (Akobundu, 1987).

12.9 Nematode and Borer Control

Nematode and borer control is vital in banana production. They destroy the root system and insert tunnels in the combs causing the plant to be extremely vulnerable to any wind storm. Nematicides commonly used include: nemacur, curlone, rugby, miral, temik, furadan, and counter. Temik has been banned, so it is no longer in use anywhere in Cameroon. Safety regulations were recommended but not strictly followed.

12.10 Fertilisation

Section 3.2.2 covers the macronutrients generally required by banana production. On average, approximately 600 kg K/ha/yr and 400 kg N/ha/yr are applied. Dolomite is rarely used by the semi-modem plantations in Cameroon, even though some expressed intentions to do so. Fertiliser application is usually done manually using a pre-calibrated applicator. The base of the plant is kept clean before application. Application is done in a semi-cylinder shape on the daughter side of the plant approximately 30 to 40 cm away from the base of the mother plant.

12.11 Aerial Spray

Aerial spray application is necessary for Sigatoka control. Small helicopters, equipped with rotary atomisers, were utilised. They were practical in the semi-modem plantations due to the uneven topographical features. Occasionally, small airplanes were also adopted. Under normal weather conditions, the aircraft could spray 200-250 ha of fungicide per hour. Ten to twelve applications/year will suffice. The following chemicals or a combination were frequently used in various concentrations: bravo, tilt, mancozeb, maneb, calixin, calixin-maneb, trinton and spraytex oil. There was a mixing station which facilitated the operation.

12.12 Cableway Maintenance

There existed no cableway network in the semi-modern production-system in Cameroon. The uneven topographical features characterised by the system was partially responsible. Other reasons included high cost of equipment, installation, and maintenance.

12.13 Harvesting

Two harvesting systems were adopted by the semi-modern plantations. About 75% of the farms adopted the eye-harvesting technique, i.e. the fruit was not calibrated prior to cutting. A "V" mark was put on the pseudostem to gradually lower the uncut bunch. The "backer" placed a metallic, rectangular-shaped basin with a foam pad inside beneath the bunch and eventually on his head. The bunch was cut off and the "backer" transported it to a tractor-driven trailer on the roadside. The distance varied from 100-200 m. The trailer handled 60 bunches maximum. The tractor hauled the fruits to the packing station for processing.

In 25% of the farms, the calibrated harvesting technique was used. If the bunch reached the recommended calibration, a "V" mark was put on the pseudostem to gradually lower it. An inflated tube, wrapped with a thick polyethylene bag, was placed underneath the bunch. The "backer" carried the bunch on his shoulder before it was cut. He then transported it to the roadside where it was stacked on padded leaves or hung on wooden posts implanted at various locations around the farm. Subsequently it was loaded on either a truck, lorry, or tractor-driven trailer, one on top of the other, with padded, foam in between and hauled to the packing station for processing.

12.14 Packing Station

Packaging is one of the most important operations in commercial banana production. It is the last and crucial stage in the production process. Successful sales largely depend on how well the product is packed. Even though banana is the freshest fruit in any supermarket around the world, it is also the most sensitive and highly perishable.

Rough packaging and handling cause several undesirable damages such as neck injuries, latex stains, box bums, bruisings, and scarrings. These damages greatly affect the general appearance and quality of bananas, as they leave dark blemishes on either green or the ripened fruits. Consequently, the blemishes distract potential customers, which subsequently reduces sales volume and leaves regular customers complaining.

12.15 Plastic Removal and Deflowering

As the tractor (from the eye-harvesting technique farms) arrived at the packing station, a worker went around cutting off the knotted portion of the polyethylene bags on the stalk with a sharp knife while they were being hung on roller conveyors at the entrance to the station. As soon as the plastic bags were removed, the fruits were immediately calibrated to separate the categories, and then deflowered.

In the other farms, the bunches arrived at the packing station without the polyethylene bags. They were immediately discharged from the tractor and hung on roller conveyors for deflowering and dehanding. There were no fruit patio in any of the semi-modern packing stations.

12.16 Bunch Sampling and Quality Inspection

No sampling or inspection was actually done.

12.17 Dehanding and Selection

After the various categories had been identified, they were dehanded with a sharp, curved knife and put in separate tanks. There was a dehanding tank for each category, i.e. CATEGORIES EXTRA, ONE and TWO, respectively. Selection with a curved-neck knife into clusters of various sizes and rejection of unwanted fingers and hands followed. The dehanding and floatation tanks were full of pressurised water, which washed and pushed the fruits to the trayers, simultaneously. These tanks varied in sizes. In the semi-modern plantations, the dimensions range from 2 metres by 1 metre to 3 metres by 2 metres. The duration of the clusters in the floatation or delatexing tanks

106

varied from 1-5 minutes, thus leaving most fruits still oozing out latex. The oozing latex later deteriorated the quality of the fruit especially while in transit.

12.18 Weighing and Labelling

Weighing was randomly done after packing, but never prior to packing. Labelling did not exist but arrangements for future adoption are underway.

12.19 Spraying

Spraying of banana clusters was manually and crudely done. The crowns of clusters from the delatexing tanks were simply steeped into a basin full of premixed fungicides to prevent post harvest rot, moulds, and disease infestation in transit. Fungicides commonly included tecto, benlate, and raertect. With this method, it was difficult to maintain a consistent chemical concentration. Knapsacks were utilised for crown spraying in 25% of die semi-modern packing stations.

12.20 Packaging

Since the target market for fruits from the semi-modem farms are the French markets, two types of boxes - 20 kg and 17 kg - were used. About 75% of the farms have standardised all categories into the 20 kg boxes whilst others still segregate, i.e. the EXTRA CATEGORY was packed into the 20 kg box, whilst CATEGORIES ONE and TWO were put into the 17 kg boxes.

Furthermore, two types of packing patterns were adopted. The four-row pack with two rows at die bottoms and two on the top, and the five-row pack with three rows at the bottom and two on the top. The former was used for category extra fruits while the latter was for CATEGORIES ONE and TWO, if packed in the 20 kg boxes. The four row pack was adopted for 17 kg boxes. A complete box was made up of a body, a cover, a flexible pad, and a plastic liner. The tare of a box ranged from 1.1-1.6 kg.

Vacuuming and palletising were not done in the semi-modern system. Also, quality inspection at the port was not carried out in this system.

12.21 Packing Station Manpower Requirement

An average of 90 persons could produce approximately 2,500 boxes/packing station in a ten-hour working day. However, the boxes/man/day was between two to three in a semi-modern set up.

Chapter Thirteen

Modern Production Technology

13.1 Land Preparation

Contrary to the semi-modem system, the modem plantation is characterised by a consistent and even topography. Since the plantation is on a permanent basis, it is generally classified as "virgin" land. A soil survey and maps, illustrating soil class, planting radius, cableway and drainage directions, roads, offices and housing facilities, were drawn. Other major operations included felling of trees with chain saws, piling, and burning. This was followed by handslashing of shrubs and herbicide application for weed control. Furthermore, irrigation pipes were buried and drainage canals were constructed. Finally, staking was done and then planting followed.

13.2 Planting Material

Planting material was obtained from seed propagation as discussed in Section 12.3. The in vitro micropropagation technique was attempted in certain parts of the only existing modern plantation with great success. Furthermore, the "bud stimulation" propagation method, expanded by Hamilton (1965), and unproved by Menendez and Loor (1979), and Turner (1968), were successfully implemented. The small unwanted buds were extracted from the mother plant, slightly peeled and put into plastic bags (Stover and Simmonds, 1987). They were conveyed to a nursery and eventually transplanted after 2-3 months. Experiments with water suckers also turned out successful.

13.3 Planting

Initially the double row planting method was adopted. The spacing was 1.125 metres by 5 metres with a plant population of 1,777 mat/ha. The hexagonal or conventional planting pattern, with a spacing of 1.6 metres by 3.3 metres and plant population of approximately 1,900 mat/ha was implemented in some parts of the plantation. Furthermore, the double-row section was gradually being transformed into the conventional pattern. Based on the total plant population of double-row and the conventional pattern, the latter maximises space more than the former.

Planting equipment included: (a) a wire string that would not stretch or cut when pulled about 50-100 metres long (the planting distance was already labelled on the string); (b) a stake attached to each end of the wire string; (c) two stringmen, one on each baseline; (d) additional stakes to be implanted in the recommended aisle distance, 3.3 metres in this case; (e) the planting crew; and (i) the seed distributors.

The seeds, usually maiden suckers, were peeled and cut short, and only the rhizome part was planted. Bullheads were occasionally used in times of shortages. The string was tensioned by the stringmen. Holes were dug, commensurate with the size of the sucker, by the planting crew, using shovels at the labelled planting spot and distance on the string. The pre-distributed seed was buried in the hole after the top layer of soil had been put back. The bottom layer was used to complete the job. To ensure soil firmness and seed immobility and to prevent water saturation, die planter stepped on the planted seed. The procedure continued until the operation was completed.

13.4 Pruning

In the modern plantation the pruner was provided with the following equipment: (a) a broad machete (bolo); (b) a sharpener; and (c) ribbons or plastic tapes for marking. The operation was in a seven to eight weeks cycle. It was either parcelised, i.e. an individual was assigned to a specific area (45 ha) and was solely responsible for that area, or centralised, i.e. group work. This operation is so delicate that if wrongly done, the whole plantation is affected. (See Section 11.4 and 12.5 for more details).

13.5 Replanting

Profit maximisation is directly correlated with the number of boxes produced. The number of boxes packed depends on the size and number of bunches harvested. Replanting, which is the replacing of missing points, guarantees a consistent plant population and thus is a necessary condition for profit maximisation in a modern production system. The operation is recommended twice a year. The primers played an important role for die success of this operation. They did the staking of missing points and identified seeds to be replanted. The replanting crew followed the pruner, desuckering marked seeds, peeling, digging holes, trimming leaves and replanting them.

13.6 Plant Population Count

Plant population count was recommended at the end of each pruning cycle. The specific objective was to obtain an even population distribution based on the soil class and growth factor. The "Mat splitting" technique was occasionally implemented in areas with inferior population density and for seed propagation. An 8 metre long cord was attached to a bar and stretched out. All the plants along that space were added up and multiplied by 50 to obtain the total production unit/ha. About seven readings/10 ha provided an accurate average count (Stover and Simmonds, 1987).

13.7 Propping, Bagging, Debudding, and Dehanding

Sections 11.5 and 11.6 discussed this operation at length. Only the twine and, from time to time, the bamboo propping methods were utilised. In the modem plantation this operation was on a weekly basis and was only undertaken when the bunch reached baggable stage, i.e. between 20-22 weeks after planting. This operation was jointly carried out with bagging, debudding, and dehanding. The following equipment was needed: (a) a wooden ladder; (b) a knife; (c) twine 12-14 metres long; (d) ribbons; and (e) a metal needle. The bagger placed the wooden ladder against the pseudostem and ascended it. The propping rope was fastened around the pseudostem underneath the stalk. The bag was put around the bunch and attached slightly above the placenta mark on the stalk. Dehanding of the false hand plus either one or two true hands was done. The bud was cut about 10 cm away from the arrester finger and a small segment of it was cut off for recording purposes. In addition, the block number was written on the index finger to identify trouble spots in future.

111

13.8 Irrigation

This plantation had an undertree irrigation network. Sprinklers were installed every 8.3 m, with approximately 12 sprinklers every 100 m. Each sprinkler line supplied water to two planting lines. The sprinkler lines (Poly Etheyne pipes or PE 16) were buried 15 cm in the ground and in the same direction as planting lines, whilst the (Poly Vinyl Chloride or PVC) pipes were buried 40 cm in the opposite direction. The problem with this system is the exposure to damage, theft, and short life span of the sprinklers.

13.9 Weed Control

Same as in Section 12.8. In addition, ring weeding/sanitation was occasionally undertaken before fertiliser application.

13.10 Nematode and Borer Control

Same as in Section 12.8. The chemicals commonly used were nemacur, furadan, temik, and counter. Temik is banned and is no longer applied.

13.11 Fertilisation

Sections 4.1.2 and 12.10 discussed fertilisation in detail. However, due to the soil type, dolomite was also used in the modem production system.

13.12 Aerial Spray

Same as in Section 12.11. However, due to the even topographical features, only small aircraft were used.

13.13 Cableway, Roller and Spacer Bar Maintenance

The cableway network not only facilitates transportation of banana bunches from the field to die packing station, but also reduces damages. The rollers were washed and greased bi-monthly using a "greasegun". There was a standby maintenance crew responsible for all repairs.

13.14 Harvesting

Only the calibrated harvesting technique was adopted in the modern production system. Harvesters were in gangs of four to seven people/gang depending on the harvesting distance. For instance, a gang of five was made up of a calibrator-cutter (the person who actually cut the bunch off the pseudostem), two backers (those who carried the bunch to the cableway), and two haulers (those who transported the bunches to the packing station). Aerial tractors were also used in this plantation.

Each gang was provided with the following equipment: (a) one machete and a stick with a fixed calliper attached to it for the cutter; (b) two shoulder pads made out of tubes, for the backers; and (c) 50-75 rollers and 48 or 72 space bars. Once a bunch that met all the required specifications was identified, the cutter inserted a "V mark on the pseudostem and slowly pulled it downward. The backer placed it on his shoulder and took it to the cableway as soon as it was cut. The bunches were immediately taken to the packing station, 25 at a time. Between 100-125 bunches were recommended for the aerial tractor.

13.15 Plastic Removal and Deflowering

Fruits coming from the field were directly taken into the fruit patio. The plastic bag was neatly torn, folded upward, and inserted between the hanging string and the stalk. Then the bunch was deflowered by the same person. A maximum of five bunches was recommended to minimise latex stain.

13.16 Bunch Sampling and Bunch Inspection

This operation provided a clear picture of what was happening in the field. It provided information such as: bunch weight, age, number of bunches hauled at a time, calibration, hand class, number of rejected bunches and quality. Two bunches were sampled from each train of 25 bunches. The information was regularly analysed and the feedback was immediately sent to the farm administrators or farm managers for corrective measures. Two people were responsible for the sampling and one for bunch inspection.

13.17 Dehanding and Selection

All the hands were put into the dehanding tank full of water, irrespective of the categories. Only the floatation tanks were segregated into CATEGORIES, EXTRA and ONE. There were special tanks for banana pack, an equivalent of category two, the 13 kg and the institutional pack (single fingers). The selectors were well trained and it was they who determined the various categories using fixed callipers. Bananza and the 13 kgs were pre-selected and put on the conveyor belt for transfer to a special tank. These clusters were reselected and sent into the category two floatation tank. Each selector was provided with a padded board, measuring tapes, clusterisation knife, fixed calliper, sponge, and detergent.

13.18 Traying and Spraying

Traying began after the fruits had been submerged in water (floatation tank) for a <u>minimum</u> of 20 minutes. The trayer checked for defects and arranged clusters in groups of small, medium, and large sizes on a three-row tray. The fruit was then sprayed with a spraygun that pumped out pre-mixed fungicides (tecto or mertect and alum) to prevent post-harvest rots, moulds, and any disease infestation in transit.

13.19 Weighing and Labelling

All the trays were weighed on a scale prior to packing and randomly weighed after packing to ascertain that the required weight was respected. Labelling (using the producer's logo) was done to differentiate the fruit from others when it reached the targeted world market. The number of labels placed on the middle inner whorl of the fingers depended on the size of the cluster. Usually a sticker was affixed alternatively on larger clusters. Only one sticker was placed on a three or four finger cluster.

13.20 Packaging

Only the 20.65 kg boxes were being used for CATEGORIES EXTRA, ONE and BANANZA. Experiments with the 18 kg palletised boxes were successfully conducted but awaited full-scale implementation. The 13 kg pack and the

114

loose finger or the institutional pack (IP) experiments were still at the initial stage, pending full-scale implementation in 1992. These were designed for the British market. The IP pack is strictly for the Sainsbury market in the United Kingdom.

Two types of packing patterns were used. The four-row pack, for category extra, and the five-row pack, for category one (Section 12.14). A radio-like divider was used-for the four-row pack.

13.21 Vacuuming

After packaging, a vacuum pump which sucked out air and water residues was inserted into the closely tightened banavac. This operation was pertinent, especially for long voyages, as it improved fruit respiration, prevented ripening, and inhibited the spread of diseases.

13.22 Palletising

Special 13 and 18 kg boxes were used for pallets. A pallet contains 48 boxes (18 kg), in a 6-by-8 loading formula and 63 (13 kg) boxes in a 7-by-9 loading formula. Either a manual or automatic striper machine was used to tie the boxes firmly. A manual fork lift (pallet jack) was used to load the pallets in a container. Six persons were required for the operation, but empirically, the number required depended on the number of lines operating daily.

13.23 Labour Requirement

Manpower requirement depended on the required output and number of lines operational. There were 12 persons/line, plus nine auxiliaries for less than three lines and 12 auxiliaries for three or more lines. All things being equal, 4,500 boxes could be packed with four lines (10 workers) in ten hours per day. On average, the boxes/man/day ranged from four to six in a modern production system.

13.24 Port Quality Inspection

In the modern production system, wharf quality evaluation played a vital role, as it provided day-to-day feedback on quality standards. Ten boxes of each category were set aside for 12-24 hours to allow for the appearance of any invisible damage sustained in the packing station. All the clusters were inspected, and records on the maximum/minimum finger length, maximum/minimum grade, various defects such as scarring, bruising, insect damage, maturity stain, etc. were kept. A quality score (Q-score) and the percentage of clusters meeting specification (PCMS) were calculated. A Q-score of 80 and above was generally accepted, but this also depended on the market and the competitiveness of other producers. A minimum inspection of one percent of the total boxes packed was required.

13.25 Summary

Chapters Eleven, Twelve, and Thirteen focused on the three different technologies adopted in banana production in Cameroon. The various systems were the traditional, semi-modem, and modern technologies, respectively. The traditional system was adopted by local farmers and the products were strictly for domestic markets. The semi-modern and modern systems were adopted by corporate farmers and their products were for both domestic and world markets. However, only the portion of the product that did not meet the world market quality specifications were channelled to the domestic market.

Chapter Fourteen

Endemic Problems in Banana Production And Marketing

Banana production is a complex, risky, and ambiguous operation. Yet, optimisation under risk and uncertainty is a prerequisite for success. The problems encountered are numerous and are spread throughout each stage of the day-to-day operation. Most of them are unpredictable but require spontaneous solution. The solution to one problem creates other problems. Finally, they become a vicious circle. In this section, only the major problems are discussed.

14.1 Input Availability

Banana production requires several inputs such as fertilisers, pesticides, herbicides, and other factors of production. Unfortunately, Cameroon is a net importer of the expensive but necessary inputs, which subsequently drive up cost of production. Furthermore, the inputs were frequently scarce when highly needed. The application of certain inputs such as potassium (K) and urea (N) require timeliness (Sections 4.1.2.1 and 4.1.2.2) for a successful and maximum plant uptake. Therefore, wrong timing simply means mismanagement of resources as the inputs and manpower are wasted, thus increasing production costs.

14.2 Pests and Diseases

The presence of major pests (Chapter Five), such as the banana borer *(Cosmopolites sordidus)*, banana nematodes *(Radopholus similis)*, banana rust thrips *(Chaetanaphothrips orchidii)*, banana aphid *(Pentalonia nigronervosa)*, banana peel scarring beetles *(Colapsis spp.)*, banana peel-feeding caterpillar *(Platynota rostrana* or *Ecpanteria spp.)*, the giant African land snails and other smaller species *(Achatina fulica* and *Limicolaria aurora)* etc., requires immediate control measures if the population is greater than the economic level.

Control measures vary from cultural and biological to chemical. A combination of these measures usually proved more effective, whereas cultural and biological measures were less effective. Once more, timeliness is very important. The unavailability of a given chemical, because of the time required for importation, may be just enough to have the whole plantation destroyed. For instance, the African snails *(Achatina fitlica, Limicolaria aurora, tic.)* attacked almost all of the Cameroon banana plantations in 1991 with a serious repercussion on the total exportable output, quality and profitability. Unfortunately, very little was known about the pest and thus no control device was actually implemented. Ever since the substantial loss - approximately 50% in the modern system - extensive studies have been carried out. A combination of mechanical, physical, and chemical control methods are the recommended possible solution (Adamu and Fonsah, 1992).

On the other hand, diseases such as black Sigatoka *(Mycosphaerel/Ut fijiensis)*, Panama disease *(Fusarium ozysporum f. Cubense)*, Moko *(Pseudomonas Xanthomonas Solanaceanun)*, Cigar-end and some physiological disorders such as maturity stain, red rust, fused fingers, malformed fingers and thrips damage are major problems to banana producers. Some of these physiological disorders, such as maturity stain and red rust are seasonal. Similar to pests, they require either cultural, chemical, or a combination of these control measures, which are usually very expensive. These diseases/viruses are highly destructive if not treated on time. Successful treatment requires well trained, qualified staff, and chemical availability.

14.3 Wind Damage

Wind damage is a major problem to banana producers all over the world. Wind speed above 40 km/h can completely destroy a plantation (Stover and Simmonds, 1987). In 1960 and 1964, 1,600,000 plants were destroyed in the Cameroon Development Corporation (CDC) plantation (Lecoq, 1972). Recent observation shows that an average loss of more than 200,000 plants were sustained by the Cameroon banana growers in 1991 and 1992, respectively.

Wind damage (blow downs) is one of those uncontrollable variables associated with banana production. Others include flood and hurricanes. Wind damage has both long and short-run repercussions in terms of costs and profitability.

14.4 Maintenance of Equipment

None of the machines adopted in banana production is locally manufactured. Consequently, the problem of spare pans arises. When unavailable, these have to be imported and will take time to arrive. When available, an expert is needed to get the job done. When all the costs incurred in repairing a piece of equipment are calculated, one finds out that it is very expensive. As a result, production costs are escalated. Occasionally, if a spare part is not found abroad, the expensive piece of equipment is rendered obsolete as the salvage value significantly reduces, especially if it is one of the "asset fixity" equipment (Johnson, 1958), which has less value in a non-farming-environment.

14.5 Stevedoring Operation

Rough handling of the fruits at both the port of departure and arrival was a major problem and has a direct impact on the overall quality. Consequently, about 5-10% of the fruits were either discarded and/or repackaged, thus rendering the market equilibrium price for the portion that withstood the shock very low as a result of degrading quality. To minimise this problem, palletisation was recommended even though the creation of unemployment in other sectors of the economy was envisaged.

14.6 Low Prices

Cameroon producers have always obtained the lowest prices for their products, whereas the cost of production has always been at its maximum. Consequently, net returns have frequently been negative, or slightly positive after the Government injected subsidies (FAO, 1986).

14.7 Mismanagement

A major set-back of the firms under the three different technologies in this study (traditional, semi-modern, and modern) was mismanagement. It was revealed that policy implementation was determined by the personality of the company strategists rather than by "purely economic" factors. In certain instances, the strategists were technically efficient but their approach to problems were managerially inefficient, and vice versa. As a result, adopted decisions were not backed up by coherent assessment, to the detriment of die companies.

Other deficiencies included the unwillingness to respect peer values in decision making. This resulted in communication break-downs at the managerial level, even though a collective action was absolutely necessary for a successful policy implementation. This problem was more prominent in the modern than the semi-modern system, and indeed provoked a lot of ill feeling and a lack of cooperation among the management staff. Besides adopting suppressive measures and stretching the subordinates to the limit, no attempts were made to partially satisfy their needs and values. Instead, every effort was made to maximise personal, rather than the company's wealth, and to acquire more power. Both the semi-modern and modem systems practised this.

These top management egoistic and idiosyncratic attitudes partially contributed to the demise of several banana firms, especially the semi-modem, as the subordinate's loyalty was lost, thus making it impossible to get the job done. Christensen et al (1982) recommended the following for a proper strategic formulation:

(1) Appraisal of present and foreseeable opportunity and risk in the company's environment;
(2) Assessment of the firm's unique combination of present and potential corporate resources or competences;
(3) Determination of the non-economic personal and organisational preferences to be satisfied; *and*
(4) Identification and acceptance of the social responsibilities of the firm.

The strategic decision is one that can be reached only after all these factors have been considered and the action implications of each assessed".

Furthermore, and according to Christensen et al (1982),

"When a management group is locked in disagreement, the presence of power and the need for its exercise conditions the dialogue. There are circumstances when the exercise of leadership must transcend disagreement that cannot be resolved by discussion. Subordinates, making the best of the inevitable, must accept a follower role. When leadership becomes irresponsible and dominates subordinated participation without reason, it is usually ineffective or is deposed. Participants in strategic disagreements must not only know then- own needs and power, but those of the chief executive. Strategic planning, in the sense that power attached to values plays a role in it, is a political process. You should not warp your recommended strategy to the detriment of the company's future in order to adjust it to the personal values you hold or observe. On the other hand, you should not expect to be able to impose without risk and without expectation of eventual vindication an agreement, an unwelcome pattern of purposes and policies on the people in charge of a corporation or responsible for achieving results. Strategy is a human construction; it must ultimately inspire commitment. It must stir an organisation to successful striving against competitions. Some people have to have their hearts in it".

14.8 Finance

Financial constraint was a major problem in all the three production systems in the Cameroon banana industry. As earlier stated, banana production requires huge amounts of capital assets and a cash flow. All the firms had limited working capital, thus making them unable to settle their short-term debts and the resultant early maturity of their long-term debts. An increase in export volume or a projected expansion, for instance, would require a concomitant increase in several input variables, such as labour, planting material, fertiliser, herbicides, nematicides, transportation, etc., which must be financed.

14.9 Skilled Workers

Banana production in the semi-modem and modern systems require highly trained personnel, especially at the managerial level. Past information depicted that an unskilled and uneducated labour force was adopted in all operations, ranging from general labour to the staff level. The only skilled labourers were the experts in the modem system, who had an average of twenty years experience in the banana industry. Conversely, most of the experts in the semi-modern system were graduate agriculturalists, with little or no prior experience in banana production. They were being trained "on the job" and yet enjoyed the privileges of the experts.

However, the experts in both the modern and semi-modern systems were confronted with various constraints such as language barriers, cultural constraint, and the inability to transfer technical know-how to the local staff who were educationally not prepared to absorb advanced banana technology. Recent trends, however, showed an improvement in the skilled labour force, as many Cameroonian agronomists, agricultural economists, and engineers were being hired to replace the older ones. The advantage of the replacement policy is that it takes a shorter time for the younger Cameroonians to master the complexities and uncertainties involved in banana production due to then- stronger academic background.

14.10 Unreliability of Workers

Workers in modem and semi-modern technological systems were very unreliable. A successful policy implementation required scrupulous supervision. If not closely monitored, a worker assigned to apply inputs - i.e. fertiliser, nematicide or weedicide - would deliberately conceal some bags of fertiliser or litres of chemical to either sell after work for an additional income, or personally use it in his/her garden, or as simple sabotage in order to reduce his/her daily accomplished task. But he/she would turn in the hours equivalent to the estimated task, i.e. the ha to be covered and quantity to be utilised. This problem was observed in both the modern and semi-modem systems, respectively.

Information or data collection for future strategic planning were usually distorted. Furthermore, there was always a communication breakdown at each level. The standard operation procedure (SOP) for most field operations were not respected for any of the following reasons:

(A) The assignment was not well explained by the head men or overseers due to a communication breakdown;

(B) No appropriate follow up was carried out after assigning the

workers;

(C) The worker deliberately failed to carry out the SOP and invented an easier but incorrect way of getting the job done;

(D) The worker simply rushed through the assignment in order to generate additional income elsewhere, since salaries were very low compared to the services they rendered to the company;

(E) The workers simply wanted to get even with the unaccepted company policies by indirect sabotage; *and/or*

(F) The workers were simply resistant to new ideas or changes.

14.11 High Worker Turnover

The cost of hiring experts contributed to the high cost of production. An average of 20 and 8 experts were identified in the semi-modern and modern production systems, respectively. In both systems, the experts held 80% of the staff and junior staff positions. Besides their incredibly high salary scales and fringe benefits, they were also provided luxuriously and furnished accommodation with swimming pools, which were even more expensive than the salary of a Cameroonian staff. They were provided better cars than their local counterparts. The experts had drivers, night and day security guards, and maids (which Cameroonian staff did not have), and were paid in foreign currencies. The Cameroonian staff were deprived of all these privileges, even though they carried out equal or similar responsibilities with equivalent or better academic backgrounds. These costs could be minimised and the profit maximised if the local staff were trained and reasonably paid to do the job.

14.12 Quality

During the semi-modern and traditional era, quality was indeed inferior compared to bananas from other parts of the world. This, however, has changed since 1988, when modern technology was fully introduced into the Cameroon banana sub-sector, as it boosted competition amongst the Cameroonian producers.

14.13 Low Yields

The agricultural practices in the semi-modern and traditional production systems, such as lack of irrigation, insufficient treatment of pests and diseases, and inconsistent and inadequate application of certain inputs (macro and micro nutrients), contributed to low yields and reduced the profit margin.

14.14 Infrastructures

There is a great need for modern banana ripening and storage facilities, processing firms, and the development of large and rural modern markets for bananas and other agricultural products. Market communication network, information systems and extension services, and good roads are non-existent. These are serious problems to the Cameroon banana arena in particular, and to the Cameroon agricultural marketing industry at large.

14.15 Summary

Generally, the main problems encountered by Cameroon banana producers were lack of and/or timeliness of inputs, pests and diseases, wind damage, maintenance of equipment, stevedoring, low prices, mismanagement, finance, skilled workers, unreliability of workers, high worker turnover, quality, infrastructure, low yields and government policies (especially taxation policies).

Part Five

Profitability Analysis

Chapter Fifteen

Capital Budgeting

15.1 Cost and Returns under Traditional Technology

The costs and returns associated with the traditional production system are presented in Table 15.1 and were calculated over a ten-year period. The major operating costs were land preparation, planting materials, labour, and transportation. There was no cash inflow for year zero (Y_0) because there was no harvesting, and thus no sales. Initial cash outflow was 240,000 FCFA. Total production for year one (Y_1) amounted to 8 mts. The local market price per kg (P_d) of banana was 38.0 FCFA, thus generating a gross receipt of 304,000 FCFA against a total cost outflow of 126,000 FCFA. As a result, the net inflow for Y_1 was 178,000 FCFA.

From year two (Y_2) through year ten (Y_{10}), a 50% drop in yield was assumed due to drought, various diseases, and the lack of fertiliser application. Labour cost, transportation cost, and miscellaneous expenditures were also reduced by 50%. The net present value (NPV), i.e. the present value of net cash inflows minus the initial investment, over a ten-year period was 33,379 FCFA. Based on the "accept-reject" decision criterion, this project should be accepted. The internal rate of return (IRR) was 24.16%. Based on the "accept-reject" decision criterion, the traditional production system should be accepted because the IRR was greater than the cost of capital (r) which was 14%. The pay back period was estimated at 1.70 years. The data for years zero and one (Y_0 and Y_1) were actual, whereas years two through ten were projections.

Table 15.1 Net cash inflow of banana produced per ha under traditional technology (francs CFA)

Item		Years										
	Y_0	Y_1	Y_2	Y_3	Y_4	Y_5	Y_6	Y_7	Y_8	Y_9	Y_{10}
Cash inflow:												
Gross receipt[14]	0	304000	152000	76000	38000	19000	9500	4750	2375	1188	594
Cash outflow:												
Operating cost:												
Land preparation[15]	15 000	0	0	0	0	0	0	0	0	0	0
Planting materials[16]	100 000	0	0	0	0	0	0	0	0	0	0
Fertilisers*	0	0	0	0	0	0	0	0	0	0	0
Herbicides/ nematicide*	0	0	0	0	0	0	0	0	0	0	0
Aerial spray*	0	0	0	0	0	0	0	0	0	0	0
Bagging/propping*	0	0	0	0	0	0	0	0	0	0	0
Labour*	100 000	100000	50000	25000	12500	6250	3125	1563	781	391	195
Cableway maintenance[17]	0	0	0	0	0	0	0	0	0	0	0

[14] Gross receipt for year zero (Y_o) was zero because due to the cultural practices, harvesting did not occur until Y_1 even though planting began in April. The price/kg was 38 FCFA. A maximum of 8 mt/kg were assumed harvested in Y_1 with a 50% decrease in subsequent years until Y_{10} due to drought, pest, and diseases.

[15] Cost of hired labour for felling (chain saw), slashing, burning, etc.

[16] A sword sucker was estimated at 50 FCFA each.

Table 15.1 (Cont). Net Cash inflow of bananas produced per ha under traditional technology (francs CFA)

Item	Years										
	Y_0	Y_1	Y_2	Y_3	Y_4	Y_5	Y_6	Y_7	Y_8	Y_9	Y_{10}
Irrigation maintenance *	0	0	0	0	0	0	0	0	0	0	0
Fuel/oil/lubricant *	0	0	0	0	0	0	0	0	0	0	0
Packaging material *	0	0	0	0	0	0	0	0	0	0	0
Transportation *	15000	16000	8000	4000	2000	1000	500	250	125	63	31
Freight	0	0	0	0	0	0	0	0	0	0	0
Stevedoring *	0	0	0	0	0	0	0	0	0	0	0
Machine Maintenance *	0	0	0	0	0	0	0	0	0	0	0
Capital cost	0	0	0	0	0	0	0	0	0	0	0
Interest *	0	0	0	0	0	0	0	0	0	0	0
Others *	0	0	0	0	0	0	0	0	0	0	0
Miscelaneous *	10000	10000	5000	2500	1250	625	313	156	78	39	20
Research and Development	0	0	0	0	0	0	0	0	0	0	0
Building *	0	0	0	0	0	0	0	0	0	0	0
Mobile Equipment *	0	0	0	0	0	0	0	0	0	0	0
Cableway *	0	0	0	0	0	0	0	0	0	0	0
Irrigation *	0	0	0	0	0	0	0	0	0	0	0

Plant and machinery *	0	240 000	0	0	0	0	0	0	0	0	0
Total Outflow	240 000	126 000	63 000	31 500	15 750	7 875	3 938	1 969	984	492	246
Net Cash Inflow	-240 000	178 000	89 000	44 500	22 250	11 125	5 563	2 781	1 391	695	348

IRR $= 24.16\%$; NPV $= 33,379$; Rate $= 14.00\%$; PBP $= 1.70$ years; *Source:* Fonsah E.G. (1991) Field data.

15.2 Cost and Returns under Semi-Modern Technology

The costs and returns associated with the semi-modern production system are presented in Table 15.2 and were calculated over a ten-year period. Due to the cultural practices, a 25% harvest, equivalent to 4.25 mts, was expected in year zero (Y_0) with a gross receipt of 1,050,000 FCFA (1,028,500 FCFA plus 21,500 FCFA from the sale of rejects). An average of 242 FCFA/kg was used for the computation throughout the ten years. P_f represents the foreign price or the price of Cameroon banana in the importing country. A negative net cash inflow of 2,759,698 FCFA was obtained by subtracting the total cash outflow of 3,809,698 FCFA from the gross receipt of 1,050,000 FCFA.

The data for years zero (Y_0) and one (Y_1) were actual, whereas years two (Y_2) through ten (Y_{10}) were projections. The gross receipt for year one was based on a 100% plant crop harvest, equivalent to 17 mt/ha. A positive net cash inflow of 194,427 FCFA was realised in year one. Yields were expected to increase by ten, seven, five, and three percent in years two, three, four, and five, respectively; then level off from years six to ten. The major cost items were land preparation, planting materials (year zero only), labour, bagging and propping, and freight. Fixed cost items included building, mobile equipment, plant and machinery. A salvage value of 250,000 FCFA and 125,000 FCFA for building, plant and machinery were recovered in year ten (Y_{10}) since these items had a lifespan of twenty years.

A net present value (NPV) of -164,971 FCFA showed that the project was not viable after a ten-year period. The internal rate of return (IRR) was 12.58%, the cost of capital was 14% and the payback period was 6.35 years. Based on the "accept-reject" decision criterion this project should be rejected because the present value of net cash inflows minus the initial investment is negative. Furthermore, the internal rate of return (IRR) is less than the cost of capital.

However, from a policy and welfare standpoint, management could strive to increase yields and cut back on certain cost items such as labour, and also seek better prices for certain variable input as fertilisers, herbicides/nematicides, bagging and propping materials, packaging materials, and aerial spray chemicals. These policies might revive the firm and improve the NPV and IRR. On the other hand, if die firm is shut down, many people would lose their jobs.

Approximately 10,000 workers were being hired by the banana industry. These employees have several dependants, wives, and children. If the company closes down, they will lose their medical insurance, family allowance, pension, etc. With the present economic crisis in Cameroon, the crime rate is bound to increase, thus making the country a living hell. Consumption will decrease due to a reduction in disposable income, thus reducing the standard of living. Foreign earnings from the banana economy will be zero and Cameroon might become a net importer. Trade balances will reduce and other sectors of the economy will be negatively affected as economic activities will be limited.

Table 15.2 Net cash inflow of banana produced per ha under semi-modern technology (francs CFA)

Item	Years										
	Y_0	Y_1	Y_2	Y_3	Y_4	Y_5	Y_6	Y_7	Y_8	Y_9	Y_{10}
Cash inflow:											
Gross receipt[20]	1050000	4200000	4620000	493400	5190570	5346287	5399750	5453747	5508285	5563368	5619001
Cash outflow:											
Operating cost:											
Land preparation[21]	250000	0	0	0	0	0	0	0	0	0	0
Planting materials[22]	300000	0	0	0	0	0	0	0	0	0	0
Fertilisers[23]	70000	140000	140000	140000	140000	140000	140000	140000	140000	140000	140000

[20] The gross receipt for year zero was calculated based on 25% plant crop harvest. Year one was based on 100% plant crop harvest. Year two was 50% plant crop and 50% ratoon. Years five to 10 is 100% ratoon. An average of 17 mts/ha was produced in Year one and 4.25 mt in Year zero. A box of 20.65 kg banana was estimated at 5000 FCFA irrespective of the category. Revenue from the sale of rejects is included. Export volume increased by 10, 7, 5, and 3 percent from Year two through Year five and by 1% frm Year six to Year 10.

[21] Includes cost of felling, piling, slashing, burning, staking, survey, etc.

134

22 The cost of 2000 suckers was estimated at 150 FCFA each. This includes selling price, transportation, packing, hauling, planting, treatment and labour.

Table 15.2 (Cont). Net cash inflow of banana produced per ha under semi-modern technology (francs CFA)

Item	Years										
	Y_0	Y_1	Y_2	Y_3	Y_4	Y_5	Y_6	Y_7	Y_8	Y_9	Y_{10}
Herbicides/nematicide[24]	150000	300000	300000	300000	300000	300000	300000	300000	300000	300000	300000
Aerial spray[25]	75 000	150000	150000	150000	150000	150000	150000	150000	150000	150000	150000
Bagging/propping[26]	100000	200000	220000	235400	247170	254585	257131	259702	262299	264922	267571
Labour[27]	1260000	1260000	1386000	1483020	1557171	1603886	1619925	1636124	1652485	1669010	1685700
Cableway maintenance[28]	0	0	0	0	0	0	0	0	0	0	0
Irrigation maintenance[29]	0	0	0	0	0	0	0	0	0	0	0
Fuel/oil/lubricant[30]	15000	30000	33000	35310	37076	38188	38570	38955	39345	39738	40136

[23] Only nitrogen and potassium were calculated. The semi-modern system under study never used dolomite, so cost was excluded. Only 50% of the yearly fertility was applied on Year 0 since planting actually commenced in April. The stated amount was the average from all semi-modern systems under study. Fertiliser cost was constant from your one through ten.

[24] In year zero only 50% chemical weeding and nematicide control was done. Chemicals used included basta, Armada, temik, counter, and furadan. The amount is the given average of all the semi-modern systems.

[25] Five aerial spray applications were done in year zero. The calculation included cost of chemicals aircraft rental, flying time, plus trimming, fueling, and the pilot's fee.

[26] This includes the perforated bags, twines, needles and ribbons. It is assumed that 50% of the plants reached baggable stage in year zero but only 25% was actually harvested.

136

[27] Labour costs for years zero and one is the average of the total cost of labour directly contributing to the production of bananas divided by the number of hectares. The costs in Y_2 through Y_{10} is based on the assumed percentage change in production.

[28] It is assumed that there is no cableway network, thus maintenance is zero.

[29] It is assumed that there is no irrigation installation, thus zero maintenance cost.

Table 15.2 (Cont). Net cash inflow of banana produced per ha under semi-modern technology (francs CFA)

Item						Years					
	Y_0	Y_1	Y_2	Y_3	Y_4	Y_5	Y_6	Y_7	Y_8	Y_9	Y_{10}
Packaging material[31]	62500	250000	275000	294250	308963	318231	321414	324628	327874	331153	334464
Transportation[32]	26250	10500	115000	123585	129764	133657	134994	136344	137707	139084	140475
Freight[33]	243750	975000	1072000	1147575	1204954	1241102	1253513	1266049	1278709	1291496	1304411
Stevedoring[34]	14875	59500	65450	70032	73533	75739	76496	77261	78034	78814	79603
Machine maintenance[35]	6250	25000	27500	29425	30896	31823	32141	32463	32787	33115	33446
Capital cost											
Interest[36]	161073	161703	154190	146076	136509	125230	111932	96255	77772	55980	30289
											30289

137

[30] This cost is obtained by dividing the total cost of fuel/oil/lubricant used on equipment directly contributing to production/year by the number of hectares. It is assumed that 50% was used in Y_0 and 100% in Y_1. The amount for Y_2 through Y_{10} is based on the percentage change in production.

[31] This is the total cost of packing material/year divided by the number of hectares. It includes the cost of a complete empty box, i.e. bottom, top, linen, and pad, cost of reassembling, and cost of traded items indirectly contributing to packaging e.g. clusterisation and dehandling knives, and crown spraying chemicals, e.g. alum, tecto, mertect, or benlate. Note that 25% of the cost of Y_1 was absorbed in Y_0. The subsequent years were based on the percentage change in total exportable yield.

[32] The average transportation cost/mt was 6,176 FCFA for the semi-modern system and 4300 FCFA for the modern (i.e. from the farm to the port using a rented container). For instance, 6,176 x 4.25 = 26,248, rounded to 26,500 FCFA which is shown in Y_0.

[33] Freight rate was estimated at 57,353 FCFA/mt.

[34] Stevedoring rate was estimated at 3500 FCFA/mt.

[35] Machine maintenance cost in Y_0 is 25% of Y_1 since the machines were still new and required less maintenance. From Y_1 to Y_{10} maintenance cost increased following the growth in export volume.

[36] Interest rate was estimated at 18% of total fixed cost with two years grace periods, i.e., Y_0 and Y_1, respectively.

138

Table 15.2 (Cont). Net cash inflow of banana produced per ha under semi-modern technology (francs CFA)

Item	Years										
	Y_0	Y_1	Y_2	Y_3	Y_4	Y_5	Y_6	Y_7	Y_8	Y_9	Y_{10}
Others[37]	125000	250000	250000	250000	250000	250000	250000	250000	250000	250000	250000
Miscellaneous[38]	50000	100000	100000	100000	100000	100000	100000	100000	100000	100000	100000
Research/Dev.[39]	0	0	0	0	0	0	0	0	0	0	0
Building[40]	500000	0	0	0	0	0	0	0	0	0	(250000)
Mobile equipment[41]	150000	0	0	0	0	0	150000	0	0	0	0
Cableway[42]	0	0	0	0	0	0	0	0	0	0	0
Irrigation[43]	0	0	0	0	0	0	0	0	0	0	0
Plant and machinery[44]	250000	0	0	0	0	0	0	0	0	0	(125000)

---

[37] Included loading/shipping charges such as Customs, 30,000 FCFA/shipment, phytosanitary 50,000 FCFA/shipment, marketing board 170,000 FCFA. Note that only 50% allocated for Y_0.

[38] Included such items as road maintenance (grading) and compaction, bride maintenance, and some unforeseen expenditures.

[39] Research and development was zero.

[40] Referred to office and other buildings. 250,000 FCFA was recovered in Y_{10} because only 50% of the lifespan was utilised.

[41] Referred to vehicles, motorcycles, and bicycles.

[42] Cost of cableway maintenance.

[43] Cost of irrigation maintenance.

[44] Refers to electrical generators, mixing station, weather stations, etc. 125,000 FCFA was recovered in year ten because only 50% of the lifespan of the plant was utilised.

Table 15.2 (Cont). Net cash inflow of banana produced per ha under semi-modern technology (francs CFA)

Item					Years						
	Y_0	Y_1	Y_2	Y_3	Y_4	Y_5	Y_6	Y_7	Y_8	Y_9	Y_{10}
Total Outflow	3809698	4005573	4289140	4504672	4666035	4762442	4936117	4807781	4827013	4843314	4481096
Net Cash Inflow	(27596698)	194427	330860	438728	524535	583845	463633	645967	681272	720054	1137905

IRR = 12.58%; NPV = (164,971); Rate = 14.00%; PBP = 6.35 years.

Source: Fonsah E.G. (1991) Field data.

15.3 Costs and Returns under Modern Technology

The costs and returns per ha of bananas produced in the modern production system are presented in Table 15.3 and were calculated over a ten-year period. Due to the adopted cultural practices, 25% plant crop harvest, equivalent to 8.25 mts/ha, was expected in year zero, (Y_0), thus generating a gross receipt of FCFA 1,996,500 plus 9,988 FCFA from the sales of rejects or a total of 2,006,488 FCFA. An average price (P_f) of 242 FCFA/kg was adopted for computation throughout the ten years, where P_f represented foreign price of Cameroon bananas. A negative net cash inflow of 10,134,020 FCFA was obtained by subtracting the total cash outflow of 12,140,508 FCFA from the gross receipt of 2,066,488 FCFA in year zero (Y_0).

The gross receipt for year one, (Y_1) was based on 100% plant crop harvest, equivalent to 33 mts/ha. A positive net cash inflow of 485,144 FCFA was realised. Yields were expected to increase by 20, 16, 8, and 4 percent in years two, three, four, and five, respectively, then stagnate from years 6-10. The major cost items were land preparation and planting materials (for year zero only), and labour, fertilisers, packaging materials, and freight. Fixed cost items included building, mobile equipment, cableway network, irrigation plant and machinery. A salvage value of 275,000 FCFA, 200,000 FCFA, 2,500,000 FCFA, and 125,000 FCFA for buildings, cableway, irrigation plant, and machinery were recovered, since these items had a twenty-years lifespan. The data for years zero (Y_0) and one (Y_1) were actual, whereas years two (Y_2) to ten (Y_{10}) were projected.

A net present value (NPV) of 567,788 FCFA showed that the project was viable. The internal rate of return (IRR) was 15.19%, the cost of capital (r) was 14.0% and the payback period (PBP) was 5.97 years. Based on the "accept-reject" decision criterion, this project should be accepted, because the present value of net cash inflows minus the initial investment was positive. Furthermore, the internal rate of return (IRR) was greater than the cost of capital ®).

142

Table 15.3 Net cash inflow of banana produced per ha under modern technology (francs CFA)

Item	Years										
	Y_0	Y_1	Y_2	Y_3	Y_4	Y_5	Y_6	Y_7	Y_8	Y_9	Y_{10}
Cash inflow											
Gross receipt[45]	2006488	8025952	9631142	11172125	12065895	12548531	12674016	12800756	12928764	13058052	13188632
Cash Outflow											
Operating cost											
Land preparation.[46]	1400000	0	0	0	0	0	0	0	0	0	0
Planting material [47]	300000	0	0	0	0	0	0	0	0	0	0

[45] It was assumed that only 25% (8.25 mt) of plant crop was harvest in Y_0. Gross receipts for years one was based on 33 mt/ha. Export volume increased by 20, 16, 8, and 4 percent in years Y_2, Y_3, and Y_5, respectively. A one percent increase was assumed for years Y_6 through Y_{10}. An average of 5000 FCFA per 20.65 kg box of bananas was used in computation, irrespective of the category. Year one was based on a 50% plant crop and 50% ratoon harvest. From Y_3 through Y_{10} all the plants were considered ratoon.

[46] Includes cost of felling, piling, slashing, burning, staking, survey, etc.

[47] The cost of 2000 suckers was estimated at 150 FCFA each. This includes selling price, transportation, packing, hauling, planting, treatment and labour.

Table 15.3 cont. Net cash inflow of banana produced per ha under modern technology (francs CFA)

Item	Years										
	Y_0	Y_1	Y_2	Y_3	Y_4	Y_5	Y_6	Y_7	Y_8	Y_9	Y_{10}
Fertilisers[48]	175000	350000	350000	350000	350000	350000	350000	350000	350000	350000	350000
Herbicides/ nematicide[49]	150000	300000	300000	300000	300000	300000	300000	300000	300000	300000	300000
Aerial spray[50]	75000	150000	150000	150000	150000	150000	150000	150000	150000	150000	150000
Bagging/propping[51]	117500	235000	282000	327120	353290	367421	371095	374806	378554	382340	386163
Labour[52]	120000	1200000	1440000	1670400	1804032	1876193	1894955	1913905	1933044	1952374	1971898
Cableway maintenance[53]	2000	8000	9600	11136	12027	12508	12633	12759	12887	13016	13146
Irrigation. maintenance[54]	25000	100000	120000	139200	150336	156349	157913	159492	161087	162698	164325

[48] Only cost of urea, potash and dolomite were included. 50% of the cost was absorbed in year one as planting began in April. Year one through year ten was constant.

[49] Only 50% was utilised in Y_1 and 100% from Y_2 through Y_{10}. Chemicals used included Basta, Armada, Temik, Counter, Furadan, and Roundup.

[50] Five aerial spray applications were done in year zero. The calculation included cost of chemicals aircraft rental, flying time, plus trimming, fueling, and the pilot's fee.

[51] This includes the perforated bags, twines, needles and ribbons. It is assumed that 50% of the plants reached baggable stage in year zero but only 25% was actually harvested.

144

[52] Labour costs for Y_0 and Y_1 is the average of the total cost of labour directly contributing to the production of bananas divided by the number of hectares. The costs in Y_2 through Y_{10} is based on the assumed percentage change in production.

[53] It is assumed that 25% of year one maintenance cost, equivalent to two percent of the cost of cableway, was utilised in Y_0. Cost of maintenance increased in subsequent years by the percentage increase in export volume.

[54] 25% of full cost was utilised in Y_0. From Y_2 to Y_{10} cost increased following percentage increase in export volume.

Table 15.3 cont. Net cash inflow of banana produced per ha under modern technology (francs CFA)

Item						Years					
	Y_0	Y_1	Y_2	Y_3	Y_4	Y_5	Y_6	Y_7	Y_8	Y_9	Y_{10}
Fuel/oil/lub[55]	25000	50000	60000	69600	75168	78175	78956	79746	80543	81349	82162
Packaging material[56]	250000	1000000	170280	1392000	1503360	1563494	1579129	1594921	1610870	1626979	1643248
Transport.[57]	35475	141900	2574000	197525	213327	221860	224078	226319	228582	230868	233177
Freight[58]	536250	2145000	138600	2985840	3224707	3353695	3387232	3421105	3455316	3489869	3524768
Stevedoring[59]	28875	115500	120000	160776	173638	180584	182389	184213	186055	187916	189795
Machine maintenance[60]	25000	100000		139200	150336	156349	157913	159492	161087	162698	164325

[55] This cost is obtained by dividing the total cost of fuel/oil/lubricant used on equipment directly contributing to production/year by the number of hectares. It is assumed that 50% was used in Y_0 and 100% in Y_1. The amount for Y_2 through Y_{10} is based on the percentage change in production.

[56] This is the total cost of packing material/year divided by the number of hectares. It includes the cost of a complete empty box, i.e. bottom, top, linen, and pad, cost of reassembling, and cost of traded items indirectly contributing to packaging e.g. clusterisation and dehandling knives, and crown spraying chemicals, e.g. alum, tecto, mertect, or benlate. Note that 25% of the cost of Y_1 was absorbed in Y_0. The subsequent years were based on the percentage change in total exportable yield.

[57] Transportation cost was estimated at 4300 FCFA/mt for a rented container.

[58] Freight rate was estimated at 57,353 FCFA/mt.

[59] The rate was estimated at 3500 FCFA/mt.

[60] Machine maintenance cost in Y_0 is 25% of Y_1 equivalent to 2% of the cost of the machine since they were still new and required less maintenance. From Y_2 to Y_{10} maintenance cost increased following the growth in export volume. The cost includes stitching machines, aerial tractor, plant and machinery, vacuum machines, etc.

146

Table 15.3 cont. Net cash inflow of banana produced per ha under modern technology (francs CFA)

Item	Years										
	Y_0	Y_1	Y_2	Y_3	Y_4	Y_5	Y_6	Y_7	Y_8	Y_9	Y_{10}
Capital cost											
Interest[61]	1145408	1145408	1096464	1038760	970730	890524	795963	684479	553042	398082	215389
Others[62]	125000	250000	250000	250000	250000	250000	250000	250000	250000	250000	250000
Misc.[63]	125000	250000	250000	250000	250000	250000	250000	250000	250000	250000	250000
Res./Dev.[64]	0	0	0	0	0	0	0	0	0	0	0
Building[65]	550000	0	0	0	0	0	0	0	0	0	(275000)
Mobile equipment[66]	200000	0	0	0	0	0	200000	0	0	0	0
Cableway[67]	400000	0	0	0	0	0	0	0	0	0	(200000)

[61] Interest rate was estimated at 18% of total fixed cost with two years grace period, i.e., Y_0 and Y_1, respectively.

[62] Included loading/shipping charges such as Customs, 30,000 FCFA/shipment, phytosanitary 50,000 FCFA/shipment, marketing board 170,000 FCFA. Note that only 50% allocated for Y_0.

[63] Included such items as road maintenance (grading) and compaction, bride maintenance, and some unforeseen expenditures.

[64] Research and development was excluded.

[65] The lifespan is 20 years. Since the project is designed for 10 years, 275,000 FCFA is recovered, equivalent to 50%.

[66] Referred to vehicles, motorcycles, and bicycles.

147

[67] The lifespan is 20 years. 50% of the initial cost is recovered.

Table 15.3 cont. Net cash inflow of banana produced per ha under modern technology (francs CFA)

Item	Years										
	Y_0	Y_1	Y_2	Y_3	Y_4	Y_5	Y_6	Y_7	Y_8	Y_9	Y_{10}
Irrigation[68]	5000000	0	0	0	0	0	0	0	0	0	(2500000)
Plant and machinery[69]	250000	0	0	0	0	0	0	0	0	0	(125000)
Total outflow	12140508	7540808	8510944	9431557	9930950	10157153	10342259	101111237	10061068	9988189	6788397
Net cash inflow	- 10134020	485144	1120198	1740568	2134945	2391378	2331758	2689519	7867696	3069863	6400235

IRR = 15.19%; NPV = 567,788; Rate = 14.00%; PBP = 5.97 years.

Source: Fonsah (1991) Field data.

[68] The lifespan is 20 years. 50% of the initial cost is recovered.
[69] The lifespan is 20 years. 50% of the initial cost is recovered.

However, due to the huge amount of capital investment associated with modern technology, expansion would be one of the recommended alternatives. This will reduce marginal cost (MC) and average cost (AC), respectively, but will increase output. Even though prices (P_t) will reduce if the market is saturated, profit (foreign earnings) will be maximised. If production exceeds quota requirements, the excess could be channelled to other market outlets such as Italy and Britain.

149

Management could also maximise the benefits of pecuniary economy enjoyed by large firms by purchasing the bulk of its inputs such as fertilisers, herbicides, nematicides, bagging and propping materials, packaging materials, and aerial spray chemicals, at a much reduced unit rate. Labour force should be reduced, maintaining only the competent and qualified workers. Costs can also be minimised by utilising some local materials in building structures. These policies might further improve the NPV.

Chapter Sixteen

Comparative Profitability Analysis

16.1 Comparative Costs-Returns Analysis

Table 16.1 shows an average comparative costs-returns analysis for the traditional, semi-modern, and modern banana production systems in Cameroon. Total revenue (TR) which was the total sales of bananas in both the domestic (rejects) and foreign markets (note that the traditional system was strictly limited to the domestic market), was highest in the modern system, approximately 120 million FCFA/ha over the ten years compared to about 53 million FCFA for the semi-modern system and 0.6 million FCFA for the traditional system. On the other hand, net revenue in the modern system was five times greater than the semi-modern system and 144 times greater than the traditional system.

However, the net present values (NPVs) were positive in the traditional and modern systems, 33,379 FCFA and 567,788 FCFA, respectively, compared to -164,971 FCFA for the semi-modern system. The internal rate of return (IRR) was the greatest in the traditional system. Based on the "accept-reject" decision criterion, the traditional production system is the most viable followed by the modern system. Both projects should be accepted as their NPVs were greater than zero and then: IRRs were greater than the cost of capital (r).

16.2 Input Costs Analysis

Table 16.2 shows the average input cost analysis of the three different technologies in terms of total costs over ten years. Labour utilisation was maximised in the traditional system, 61.0% of total costs compared to 33.7% for the semi-modern system and 17.96% for the modern. On the other hand, the use of fixed capital was highest for the modern system, 7.4% of total cost compared to 2.3% for the semi-modern and 5.1 % for the traditional. This means that the traditional production system is highly labour intensive whilst the modern system is highly capital intensive. The percentage of total variable cost (TVC) was highest in the semi-modern system, 97.7% followed by the traditional system, 94.9%, and modern system, 92.6%. However, total variable costs were the major cost items in the three different technologies under study.

Table 16.1 Average cost-returns analysis per ha of banana produced for the three technological systems.

Items	Traditional	Semi-modern	Modern
Total revenue (FCFA)	55,219	4,808,583	10,918,214
Total cost (FCFA)	44,705	4,539,352	9,545,734
Total variable cost	42,432	4,434,807	8,836,643
Total fixed cost	2,273	104,545	709,091
Net revenue (FCFA)	10,514	269,231	1,372,480
Total revenue/total cost (ratio)	1.24	1.06	1.14
Net revenue/total cost (ratio)	0.24	0.06	0.14
IRR (%)	24.16	12.58	15.19
Rate (R) (%)	14.00	14.00	14.00
NPV (FCFA per/ha)	33,379	-164,971	567,788
PBP (years)	1.70	6.35	5.97

Source: Fonsah E.G. (1992) Field data.

Table 16.2 Average input cost analysis in terms of total cost (TC) per ha of bananas produced.

Input factor	Traditional		Semi-modern		Modern	
	Cost	% TC	Cost	% TC	Cost	% TC
Labour	27,255	61.00	1,528,484	33.70	1,714,255	17.96
Fertilisers	-	-	133,636	2.95	334,091	3.50
Herbicides/pesticides	-	-	286,364	6.32	286,364	3.00
Aerial spray	-	-	143,182	3.16	143,182	1.50
Bagging/propping materials	-	-	233,525	5.15	325,026	3.40
Packaging materials	-	-	286,225	6.32	1,360,364	14.30
Freight	-	-	1,116,278	24.63	2,917,980	30.60
TVC	42,432	94.90	4,434,807	97.70	8,836,643	92.60
TFC	2,273	5.10	104,545	2.30	704,091	7.40
Total cost	44,705	100.00	4,539,352	100.00	9,545,734	100

Source: Fonsah (1992) Field data.

16.3 Sensitivity Analysis

Tables 16.3, 16.4, and 16.5 show sensitivity analyses per ha of bananas produced under three different technologies: traditional, semi-modern, and modern, respectively. The zero percent level represented the base case situation. The selected "most likely" values (or base case values) included sales volume, sales price, operating cost, labour cost, packaging material, aerial spray, and freight rate.

A five percent increase in sales volume increased the internal rate of return (IRR) from 24 to 30% and the net present value (NPV) from 33,379 FCFA to 53,549 FCFA, but had a slight effect on the payback period (PBP) in the traditional system (Table 16.3). On the other hand, the IRR and NPV increased from 13% and -164,971 FCFA to 19% and +627, 472 FCFA, respectively, whilst the PBP fell from six to five years in the semi-modern system (Table 16.4). Finally, the IRR and NPV increased from 15% and 567,788 FCFA to 18% and 2,075,188 FCFA, respectively, whilst the PBP fell from six years to five years (Table 16.5). Note how delicate the behaviour of the NPVs in both the semi-modem and modern systems are in response to a five percent increase in sales volume.

Tables 16.3, 16.4, and 16.5 further revealed that operating costs (total variable cost) in the three systems were more sensitive to any percentage increase. The reason for this extreme sensitivity was simply because the operating costs made up more than 90% of the total costs in each of the production systems under study (Table 16.2). Other sensitive values included freight rate, labour cost, packaging material, and aerial spray, even though the degree of sensitivity varied. This means that any policy geared at improving the firm's financial standing should be targeted towards the most volatile key variables.

Table 16.3 Sensitivity analysis per ha of banana produced under traditional production technology

Input variables	Percentage Change								
	-20%	-15%	-10%	-5%	0%	5%	10%	15%	20%
Sales Volume									
IRR	-0.6	5.66	11.87	18.03	24.16	30.24	36.29	42.3	48.27
NPV	-47301	-27131	-6961	13209	33379	53549	73719	93889	114059
PBP	>10	3.37	2.46	1.94	1.7	1.49	1.32	1.17	1.04
Sales Price									
IRR	-1.22	5.13	11.48	17.82	24.16	30.49	36.83	43.16	49.5
NPV	-49933	-29105	-8277	12551	33379	54207	75034	95862	116690
PBP	>10	3.5	2.5	1.95	1.7	1.48	1.3	1.15	1.01
Operating Cost									
IRR	53.12	44.61	37.03	30.26	24.16	18.64	13.62	9.03	4.83
NPV	102889	85511	68134	50756	33379	16001	-1376	-18754	-36132
PBP	0.97	1.11	1.3	1.49	1.7	1.91	2.29	2.77	3.58
Labour Cost									
IRR	40	35.77	31.73	27.87	24.16	20.6	17.19	13.9	10.75
NPV	78328	67090	55853	44616	33379	22141	10904	-333	-11570
PBP	1.22	1.33	1.45	1.57	1.7	1.83	1.98	2.26	2.58
Packing Material									
IRR	24.16	24.16	24.16	24.16	24.16	24.16	24.16	24.16	24.16

Source: Fonsah (1991) Field data

Table 16.4 Sensitivity analysis per ha of banana produced under semi-modern production technology

Input variables	Percentage Change								
	-20%	-15%	-10%	-5%	0%	5%	10%	15%	20%
Sales Volume									
IRR	-26.7	-12.49	-2.63	5.43	12.58	19.24	25.63	31.87	38.06
NPV	-3334743	-2542300	-1749857	-957414	-164971	627472	1419915	2212358	2004801
PBP	>10	>10	>10	8.52	6.35	4.8	3.94	3.31	2.85
Sales Price									
IRR	-96.54	-38.4	-12.7	1.47	12.58	22.52	31.99	41.33	50.74
NPV	-4949665	-3753492	-2557318	-1361145	-164971	1031202	2227376	3423550	4619723
PBP	>10	>10	>100	9.63	6.35	4.32	3.3	2.64	2.18
Operating Cost									
IRR	53.12	42.45	32.07	22.25	12.58	2.51	-9.08	-25.53	-60.35
NPV	4358899	3227931	2096964	965996	-164971	-1295939	-2426906	-3557874	-4688841
PBP	2.07	2.57	3.29	4.36	6.35	9.37	>10	>10	>10
Labour Cost									
IRR	26.46	22.93	19.46	16.02	12.58	9.12	5.599	1.96	-1.84

NPV	1436227	1035927	635628	235328	-164971	-565271	-965570	-1365869	-1766169
PBP	3.86	4.28	4.77	5.48	6.35	7.28	8.44	9.5	>10
Packing Material									
IRR	15.02	14.41	13.81	13.2	12.58	11.96	11.34	10.71	10.08
NPV	119832	48631	-22570	-93770	-164971	-236172	-307373	-378574	-449774
PBP	5.73	5.9	6.06	6.2	6.35	6.5	6.66	6.83	7.01
Aerial Spray									
IRR	13.88	13.55	13.23	12.91	12.58	12.26	11.93	11.61	11.28
NPV	-14547	-52153	-89759	-127365	-164971	-202577	-240183	-277789	-315395
PBP	6.04	6.12	6.19	6.27	6.35	6.43	6.51	6.59	6.68
Freight									
IRR	21.83	19.57	17.29	14.96	12.58	10.15	7.64	5.04	2.33
NPV	945761	668078	390395	112712	-164971	-442654	-720337	-988021	-1275704
PBP	4.41	4.75	5.17	5.75	6.35	7	7.76	8.66	9.42

Source: Fonsah (1991) Field data

Table 16.5 Sensitivity analysis per ha of banana produced under modern production technology.

Input variables	Percentage Change								
	-20%	-15%	-10%	-5%	0%	5%	10%	15%	20%
Sales Volume									
IRR	1.65	5.26	8.7	12	15.19	18.3	21.34	24.32	27.27
NPV	-5461812	-3954412	-2447012	-939612	567788	2075188	3582588	5089988	6597388
PBP	9.7	9.07	7.86	6.8	5.97	5.26	4.72	4.29	3.94
Sales Price									
IRR	-11.13	-3.38	3.37	9.49	15.19	20.63	25.89	31.03	36.12
NPV	-10073419	-7413117	-4752816	-2092514	567788	3228090	5888391	8548693	11208995
PBP	>10	>10	9.39	7.58	5.97	4.83	4.1	3.55	3.14
Operating Cost									
IRR	33.15	28.58	24.09	19.64	15.19	10.69	6.09	1.3	-3.78
NPV	9095839	6963826	4831813	2699801	567788	-1564225	-3696238	-5828250	-7960263
PBP	3.36	3.8	4.32	5	5.97	7.18	8.81	9.76	>10
Labour Cost									
IRR	18.86	17.94	17.02	16.11	15.19	14.27	13.36	12.44	11.51
NPV	2316702	1879474	1442245	1005016	567788	130559	-306669	-743898	-1181127
PBP	5.16	5.34	5.53	5.74	5.97	6.19	6.42	6.66	6.93

Input variables	Percentage Change								
	-20%	-15%	-10%	-5%	0%	5%	10%	15%	20%
Packing Material									
IRR	17.93	17.25	16.57	15.88	15.19	14.5	13.8	13.1	12.39
NPV	1893638	1562175	1230713	899250	567788	236325	-95137	-426599	-758062
PBP	5.33	5.48	5.63	5.79	5.97	6.14	6.31	6.49	6.69
Aerial Spray									
IRR	15.51	15.43	15.35	15.27	15.19	15.11	15.03	14.95	14.88
NPV	718212	680606	643000	605394	567788	530182	492576	454970	417364
PBP	5.89	5.91	5.93	5.95	5.97	5.99	6.01	6.03	6.05
Freight									
IRR	21	19.57	18.12	16.67	15.19	13.7	12.18	10.64	9.08
NPV	3411736	2700749	1969762	128775	567788	-143199	-854186	-1565173	-2276160
PBP	4.77	5.01	5.29	5.61	5.97	6.33	6.75	7.21	7.72

Source: Fonsah (1991) Field data

Figures 16.1, 16.2, and 16.3 show graphical sensitivity analyses on sales volume for the three different technologies under study. Zero percent is the base case. The steepness of the sensitivity curves indicate the degree of risk associated. In Fig. 16.1, for instance, the IRR curve of the semi-modern production system responded faster to either a positive or negative change in sales volume whereas the modern system showed a consistent growth pattern.

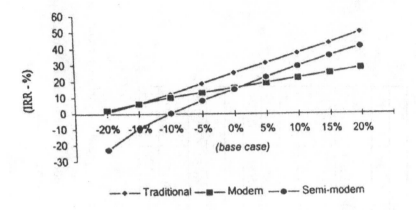

Figure 16.1 Sensitivity analysis on sales volume: a comparison of
the IRR for the three systems

Source: Fonsah (1991) Field data.

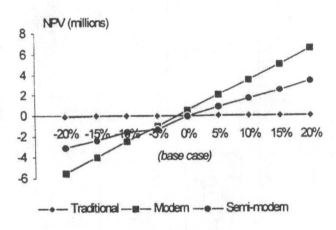

Figure 16.2 Sensitivity analysis on sales volume: a comparison of
the NPV for the three systems

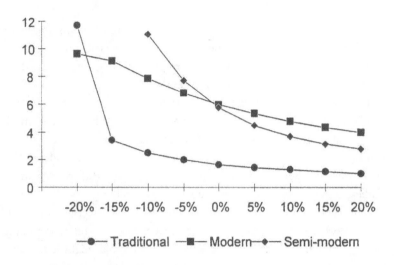

Figure 16.3 Sensitivity analysis on sales volume: a comparison of
the PBP for the three systems

Source: Fonsah (1991) Field data.

On the other hand, the NPV curve for the modern production system
was more sensitive to any change in sales volume than the semi-modern and
traditional systems, respectively. Fig. 16.2 showed that if sales volume
decreased by 20%, the NPV for the modern system would approximately
drop to -6 million FCFA, compared to approximately -3.5 million FCFA for
the semi-modern and less than -100,000 FCFA for the traditional system.

Figure 16.3 shows the sensitivity curves for the payback periods (PBP). If
sales volume dropped by 15%, it would take the traditional system slightly
above three years and the semi-modern production system more than ten
years to pay off the capital, compared to two years for the traditional and
eight years for the semi-modern system if sales volume decreased by 5%.
Conversely, the pay back period (PBP) in response to a 15% and 20%
decrease in sales volume, for the modem system, were nine and almost ten
years, respectively.

16.4 Summary

A comparative cost-return analysis of the three systems showed that the traditional and modern production systems realised positive net present values (NPV) and internal rates of return (IRR) greater than the cost of capital (r), whilst the firm using the semi-modern technology after a ten-year period had a negative NPV and its IRR was less than the cost of capital. Yields were highest in the modern system, followed by the semi-modern and traditional systems, respectively. The annual cash inflow (net revenue) was also highest in the modern system, followed by the semi-modern. The results of the semi-modern technological system were in conformity with previous studies, i.e. without government subsidies profitability would be negative.

An average input cost analysis showed that the traditional and semi-modern systems were more labour intensive, whereas the modern system was more capital intensive. Total fixed cost (TFC) was highest in the modern system, followed by the semi-modern and traditional systems, respectively. Furthermore, total variable costs (TVC) as a percent of total costs was highest in the semi-modern, followed by the traditional and finally the modern system.

A sensitivity analysis showed that a five percent increase in either sales volume or sales price would change the negative NPV of the semi-modern system to positive. On the other hand, it would improve the NPVs of both the modern and traditional system by almost fourfold and twofold, respectively. A five percent change in sales price was more sensitive to profit (NPV) than the change in sales volume because total variable costs were held constant in the former situation, whereas there was a proportionate change in total variable cost in respect to any percentage increase in sales volume. Furthermore, the internal rate of return was unproved in all the systems under study with a five or more percentage change in either sales volume or sales price. A five percent increase in sales price had a significant impact on profit because the total variable cost was unchanged.

Chapter Seventeen

Marketing Analysis

17.1 Trends in Export Volume

The data on trends in export volume, from 1978 to 1988, to France, United Kingdom, other European countries, and the average were regressed using simple linear regression model $Y = a + bt$ in which a is the y-intercept and b is the slope. The values obtained from the regressed export data are shown in Table 17.1 and Fig. 17.1. The coefficient of determination, r^2, for the United Kingdom was highest, 0.75, compared to France, 0.67, and Others, 0.15. Generally, an r^2 closer to 1 is statistically more significant because it fully illustrates how well the regression line fits the points and better explains the variation between variables Y and X (Dowdy and Wearden, 1983; Lewis-Beck, 1980). At the other extreme, a more stable slope of 0.36 was obtained for Others compared to 1.07 for France and 1.57 for me United Kingdom. The regression line equations derived were:

$$Y_{France} = 26.61 + 1.07t$$

$$Y_{UK} = -24.36 + 1.57t$$

and $Y_{Others} = -2.23 + 0.36t$

Table 17.1 Regression values for Cameroon banana exported from 1978-1988
using simple regression line Y = a + bt

Item	France	United Kingdom	Others
Mean	47.18	5.99	4.74
σ (Standard Deviation)	6.27	8.73	4.48
Intercept	26.61	-24.36	-2.23
Slope	1.07	1.57	0.36
r (Correlation coefficient)	0.818	0.867	0.387
r² (Coefficient of determination)	0.67	0.75	0.15

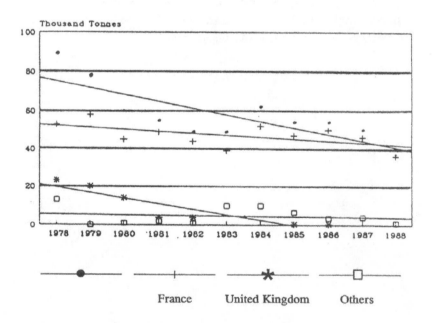

Figure 17.1 Regression Graph on Cameroon Banana Export to
France, Britain, and other European countries, 1978-88

164

Fig. 17.1 further shows a drastic downward trend in quantity exported. In 1978, 1979, and 1980, 89,000, 78,000, and 60,000 mts were exported to France, Britain, and other nations, thus reflecting a decline in export of 12.4 and 23.1%, respectively, in a two-year span (Table 17.2). Moreover, the data for 1986 and 1987 depicted an export decline of 7.4 and 26 per cent, respectively.

Further analysis illustrated that Cameroon producers never met their French market quota of 60,000 mt/yr according to the African, Caribbean and the Pacific/French overseas Departments (ACP/DOM) Lome convention requirement.

How can this downward trend be explained, when historically, Cameroon has been in the banana business for almost a century? The answer is twofold: (a) the production technology; and (b) mismanagement

Table 17.2 Percentage change in export, 1978-1991

Years	Total output (mt)	Percent in (y)	Production system
1978	89	-12.4	Semi-modern and Traditional Era
1979	78	-23.1	
1980	60	-8.3	
1981	55	-10.9	
1982	49	+1.0	
1983	49	+26.5	
1984	62	-12.9	
1985	54	+1.0	
1986	54	-7.4	
1987	50	-26.0	
1988	37		Pivotal point
1989	56	+51.4	Modern era
1990*	81	+44.6	
1991*	123	+51.8	

Source: Adapted from GIEB (1990).

* Fonsah (1992) Field data.

166

All the existing banana firms during the time period were semi-modern. Even though there were more producers then than in the 1990 era, total output could not be any better simply because of the adopted production technology. Section 6.2 dealt with the characteristics, advantages and disadvantages of a semi-modern production system.

Mismanagement and incompetent staff were major problems. Obviously, pest and disease infestation, and lack of several necessary inputs, could partially be blamed for the poor performance of the Cameroon banana firms. However, this argument is untenable since, empirically, the problem of pests and diseases can be minimised with the presence of well qualified staff, and a good farm management strategy.

Cameroon is still a virgin land as far as banana exploitation is concerned. Table 17.3 shows that as at 1991, less than 4,000 ha were being cultivated. Gross export volume was 61,720, 80,781 and 123,748 metric tonnes in 1989, 1990, and 1991, respectively. Further overall average mt/ha was 18, 22, and 31 in 1989, 1990, and 1991 reflecting a 22% increase in 1990/1989 and a 41% increase in 1991/1990 fiscal years, respectively. Del Monte was better managed in terms of mt/ha, followed by SPNP in 1991. It is worth mentioning here that from 1990, SPNP has been adopting modern production technology, such as installing irrigation systems and constructing a modern packing station. By the end of 1992, this firm will meet all requirements of a modern production system. The reported yield in 1991 compared to 1989 shows that the move towards modernisation is already bearing fruits.

167

Table 17.3 Area under banana exports and unit yield, 1989-1991

Growers	Export 1989	Area (ha)	mt/ha	Export 1990	Area (ha)	mt/ha	Export 1991	Area (ha)	mt/ha
OCB[70]	25323	1200	21	21584	1200	18	28475	1200	24
SPNP	11398	1200	9	27257	1200	23	42090	1200	35
PHP	2347	135	17	2985	135	22	3846	135	28
CDC/DMC	16578	500	33	25547	800	32	42868	1100	39
CDC/EKONA	4740	270	18	3408	270	13	5550	270	21
Others[71]	1334	90	15	26	-	-	919	90	10
Total	61720	3395	18	80781	3605	22	123748	3995	31

Source: Fonsah (1992) Field Data.

[70] OCB has been liquidated and is now called SBM.
[71] Others include Tiani, Deumo, IRA Nyombe, etc.

168

Additionally, the major problems such as Moko disease, Bunchy Top Virus (BBTV), fruit spot, Sigatoka pathogen, hurricanes and floods associated with banana production, especially in Central America and the Philippines, have not yet invaded Cameroon. Sigatoka pathogen is still in its infancy, that is why only 10-12 aerial sprays/yr are required (Foure, 1988) compared to 35-45 sprays in Central America, the Philippines, and other parts of the world (Stover and Simmonds, 1987).

A rational producer would seek to maximise profit by striving to meet its quota requirement in the French market while capturing a greater marketing niche in other unrestricted markets. This study shows that only 60% of Cameroon's quota was supplied in 1988 compared to 96.7% in 1979. In other words, a market segmentation and targeting strategy should have been adopted since consumer behaviour varied among countries, groups, and sexes.

Fig. 17.1 further shows that only 25.8% of 1978 total exportable output penetrated the British market. Thereafter, export to Britain dropped to zero in 1983, 1984, and from 1987 through 1988. Export volume to other countries improved in 1978, 1983, and 1984 as 20.4%, 16.1% and 14.6% were recorded. It gradually declined in subsequent years. An increased production would nevertheless increase export volumes, profitability, and the implementation of a geographic segmentation would be possible.

Yearly production data collected from 1978-1991 (Fig. 17.2 and Table 17.2) revealed that 1988 was the pivotal point in the history of Cameroon's banana industry. In 1989, 1990 and 1991, the gross total exportable outputs were 60,000 mts, 81,000 mts, and 123,748 mts, compared to 37,000 mts in 1988, thus reflecting an increase of 51.4%, 44.6% and 53.2%, respectively.

Figure 17.2 further depicted a decade of continuous significant decline in total output from 1978 to 1988, with 1988 being the lowest when only 37,000 mts were exported. The reasons for the drastic fall in supply, which included the adopted technology, mismanagement, pests and diseases, were discussed in Chapters Six and Fourteen. However, the creation of the newly technologically innovated banana plantation in 1988 further boosted high level competition amongst the existing producers. As a result, there was a significant progress in 1989/88, 1990/89 and 1991/90 as 62%, 35%, and 51% increases in supplies respectively were recorded compared to 13% for 1985/84 and -6% for 88/87 production years. Consequently, the increasing trend in total exportable volume can partially be attributed to the introduction of modern technology into the Cameroon banana arena.

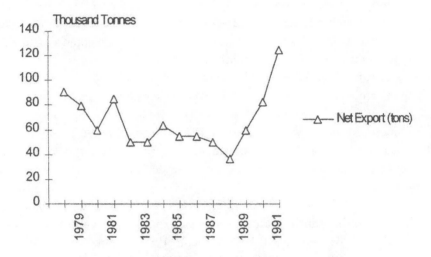

Figure 17.2 Cameroon Net Export to the World Market, 1978-91

In this book, 1989 was frequently utilised because it was the year in which all the three production systems under study were fully in place in Cameroon.

An analysis of production behaviour during the modem era, 1989-1991 (Fig. 17.3) revealed that production was at its lowest in May for 1989, July for 1990, and August for 1991. Conversely, the peak periods were November for 1989 and 1990, and December for 1991. The enormous increase in December 1991's output was because producers were advised to maximise output, since no export vessel was assigned on the first week of January 1992. Consequently, harvesting was scheduled for six weeks instead of the normal four weeks or occasional five-week schedule.

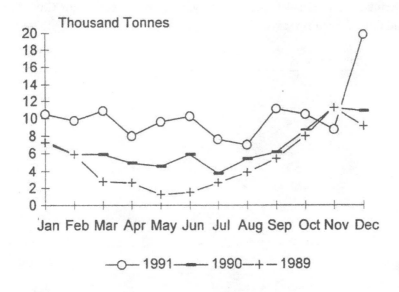

Thousand Tonnes

Jan Feb Mar Apr May Jun Jul Aug Sep Oct Nov Dec

—○— 1991 ——— 1990 —+— 1989

Figure 17.3 Monthly Export Volume from all Cameroon
Banana Producers, 1989-91

The 1989 monthly export trend for the four major suppliers of the
French preferential market (Fig. 17.4) placed Cameroon at the lowest
position in terms of volume. Martinique topped the list with a substantial
margin followed by Guadeloupe and Cote d'Ivoire. The unexpected drop in
Guadeloupe's total exportable volume in 1989 (Fig. 17.4) was due to the
hurricane which completely destroyed the plantation in September. However,
according to the ACP/DOM Lome convention (1975), the Overseas
Departments' (Martinique and Guadeloupe) market quota, was 2/3 whereas
the ACP countries were only allocated 1/3 of the total volume.

However, while the Overseas Departments have been taking advantage
of their quota privileges, the ACP countries on the contrary have not been
able to meet up in the past decades. With the newly-adopted technological
innovation, 1990 was a revolutionary era in the Cameroon banana arena as
output rose beyond the quota requirement of 60,000 mts/yr. Fig. 17.3 shows
80,785 and 123,748 mts exported in 1990 and 1991, respectively. On the
other hand, Fig. 17.4 shows the month of May as the peak period for

Martinique and Guadeloupe, and November and October for Cote d'Ivoire and Cameroon, respectively.

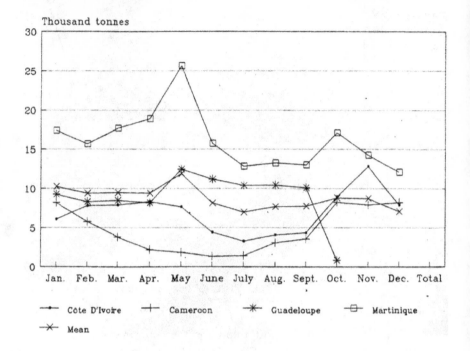

Fig. 17.4: Monthly banana export trend of four major suppliers of the French market, 1989.

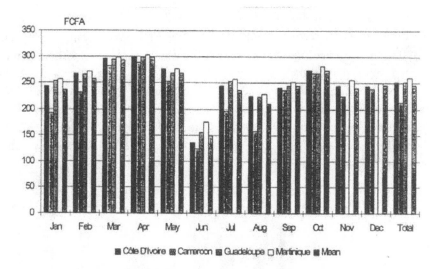

FCFA

■ Côte D'Ivoire ■ Cameroon ■ Guadeloupe □ Martinique ■ Mean

Figure 17.5 Monthly Price Trend of the Four Major Suppliers
of the French Market, 1989

17.2 Trends in Export Price (F.O.R.)

A F.O.R. monthly price trend per kg of bananas (Fig. 17.5) illustrated
that out of the four major suppliers to the French market in 1989, Cameroon
obtained the lowest price for its products. During a time of favourable prices,
i.e. April, Cameroon obtained 291 FCFA/kg compared to 302 FCFA/kg for
Cote d'Ivoire, 300 FCFA/kg for Guadeloupe, and 304 FCFA/kg for
Martinique. Conversely, Cameroon received 127 FCFA/kg compared to 137
FCFA/kg for C6te d'Ivoire, 158 FCFA/kg for Guadeloupe, and 177
FCFA/kg for Martinique during times of unfavourable prices (June).

Martinique has always received the highest prices while Cameroon has
never exceeded the mean prices of 299 FCFA/kg and 150 FCFA/kg during
the favourable and unfavourable price periods, respectively. The reason
always put forward by French officials for the price discrimination is 'poor
quality'. The 3-4 percent price differential privilege accorded to Martinique,
Guadeloupe, and Cote d'Ivoire has a significant impact on the profitability of
the Cameroon banana firms in particular, and to Cameroon balance of
payments at large.

A three-year price trend (FOR/kg), 1987-1989 illustrated in Fig. 17.6 still
places Cameroon at the bottom compared to all her counterparts from the

173

Franc zone. Yearly average prices of 203 FCFA/kg, 225 FCFA/kg, and 231 FCFA/kg were recorded for 1987, 1988, and 1989, respectively. This reflected an eleven per cent price increase for 1988/1987 and three percent for 1989/1988. On the other hand, Martinique topped the price list with 261 FCFA/kg, 270 FCFA/kg, and 262 FCFA/kg for 1987, 1988, and 1989, respectively. Martinique thus recorded a three percent increase in 1988/1987 and three percent in 1989/1988. Despite the favourable percentage increase in price, Cameroon is still at the bottom, overall, although her price position is improving.

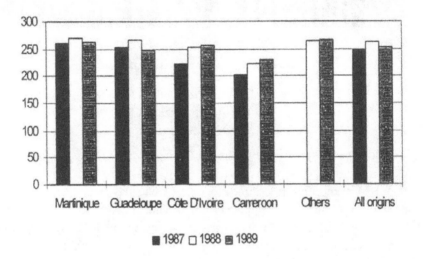

Figure 17.6 FOR Prices/kg of the Four Major Suppliers of the French Market, 1987–89

17.3 Comparison of Export Volume and Price

A comparison of 1989 monthly total export volume and Free-on-Rail (FOR) price (Fig. 17.7) illustrated that while FOR prices/10kg of bananas increased by 50% from January to April, total output for the same period decreased by 74%. Thereafter, both price and output reached their minimum in June. Subsequently, both output and prices gradually increased until October when a new peak was reached.

It is interesting to note the 1989 price and export behaviour of die four major suppliers to the French market, as illustrated in Figures 17.4 and 17.5. Martinique and Guadeloupe's timings were almost perfect as their total outputs were optimised from March to May, when prices were at their best. Cote d'Ivoire maintained a constant output, whilst Cameroon almost hit its lowest output of that year. Production and price timing are very important and are a pre-requisite for profit maximisation.

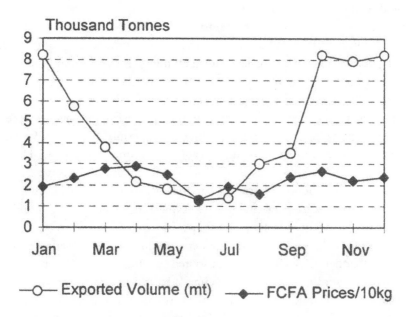

Figure 17.7 Comparison of FOR Price/10kg of Banana and Export Volume, 1989

17.4 Domestic Market and Distribution Channels

As mentioned earlier, bananas from the modern and semi-modern production systems were strictly for export/world market. Consequently, only the rejected portions of the total output penetrated the local markets.

The domestic market penetration was facilitated by some intermediaries who together formed the marketing channels.

This study revealed that three different level-channels existed in the sale of rejects from the modern and semi-modern production systems. They were: (a) the zero-level-channel (P-C), whereby the producer (P) sold the reject goods directly to the consumer (C); (b) the one-level-channel (P-R-C), whereby the producer sold to the retailer (R) who in turn resold to the final consumer; and (c) the two-level-channel (P-W-R-C), whereby the producer (P) sold to the wholesaler (W), the wholesaler resold to the retailer (R), and finally the retailer resold to the consumer (C). The zero and one-level-channels were common in the semi-modern production system in Cameroon, whereas the two-level-channel method was common in the modern production system.

Generally speaking, the Cameroon banana industry was a purely oligopolistic market, given the fact that the few existing firms produced homogeneous products. Entry was almost impossible for the following reasons: (a) an enormous amount of capital investment was required; (b) there was a quantitative restriction, a quota of 60,000 mts/yr which theoretically could be met by only a handful of producers; and (c) most of the existing firms were parastatal thus receiving heavy government subsidies in the name of "infant industry protection".

As mentioned earlier, only the modern production system adopted the two-level-channel (P-W-R-C), as illustrated in Fig. 17.8. The producer incurred no cost of transportation, loading, advertising, sales promotion, and marketing administration in the chain of distribution of the rejects from the farm to the final consumer. The wholesalers guaranteed the purchase of all the fruits from the producer. In return, they resold the rejects to the retailers who bought on the spot. The retailers provided then' own transportation and loaders. Only the marketing and general administrative costs were absorbed by the wholesalers (Table 17.4). Most of the operating costs (Table 17.5) were absorbed by the retailers. Finally, the products reached the final consumers through the retailers who had to recover all the costs incurred during the transaction and strive for profit.

Figure 17.8 Two-level-channel

On the other hand, the semi-modern production systems adopted either the zero-level-channel and/or the one-level-channel depicted in Figs. 17.9 and 17.10.

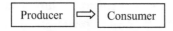

Figure 17.9 Zero-level-channel

In the zero-level-channel, the producer sold directly to the consumer without the assistance of an intermediary. The consumer in this case was usually an institution such as a school, me military, the prison, or hospitals. The purchase was done in bulk and was infrequent because existing inventory had to be cleared before repurchase was possible. Semi-modem firms adopting this method hardly maximised profits because there were few available customers. Further, it costs the producers more to carry on the role/activities of the intermediaries.

The one-level-channel (Fig. 17.10) was much better than the zero-level-channel in that the retailer usually operated as a quasi-wholesaler, thus facilitating and/or narrowing the gap between the producer and the retailer.

Figure 17.10: One-level channel

The major disadvantage was that the retailer had to get rid of the inventory before replenishment could take place. During this lag, the producer had no alternative than to throw away the surplus reject bananas. Consequently, a new policy which allowed workers to take the excess rejects for home consumption on welfare grounds had been implemented.

17.5 Marketing Costs Analysis

The marketing costs analysis in this study was strictly limited to the domestic market. In a nutshell, marketing costs included all distribution costs incurred by the intermediaries (wholesalers and retailers) from the farm gate to the consumer market. It is worth mentioning here that little or no portion of the cost was absorbed by the producers.

However, none of the Cameroon banana firms had a well established marketing department interested in the study of distribution expenses, customer groups, order sizes, and other distribution cost control units either at the national or international levels. It was further found that distribution of bananas at the international arena was solely in the hands of foreign firms such as Compagnies des Bananes, (United Brands), Agrisol, Compagnie Fruitiere, Siim, Muribane, Ste Dunand et Cie, and Simba France (FAO, 1986; GIEB, 1990), thus creating a demarcation line between the producers and the distributors.

In the modern production system, certain marketing costs were incurred by the wholesalers (Table 17.4), while consummating transactions such as: research, promotion, contact, matching, negotiation, and the facilitating functions such as physical distribution, financing, and risk were undertaken. A guarantee of all purchases had been contracted between the producer and the wholesalers. The former also guaranteed low farm gate prices of 15,000 FCFA for a tractor-load, or 5,000 FCFA/mt or 5 FCFA/kg. A tractor-load was equivalent to 3 rats. The wholesaler, in turn, resold the rejected bananas on the spot at 20,000 FCFA/tractor-load. Table 17.4 shows a wholesaler's net returns of 10,280 FCFA for a truck-load of reject bananas. An average of seven truck-loads can be sold weekly by a wholesaler. Prices are more or less fixed at this level.

Conversely, the retailers in both the semi-modern and modern systems absorbed more marketing costs as they provided their own transportation and loaders (Table 17.6), who facilitated the movement of the reject bananas from the packing station into the trucks and eventually to the targeted market in Douala. Table 17.6 also showed a net return of 69,750 FCFA and 43,500 FCFA to the retailers of the modern and semi-modem systems, respectively. Transportation in Table 17.6 was direct cost to the retailers since they were directly linked to the sale of rejects. The transportation cost incurred by the wholesaler, (Table 17.4), was indirect. This operation was very timely, fast, and important since banana is a highly perishable food commodity. Storage cost was also incurred by retailers.

Table 17.4 Income and expense statement, FCFA/truck-loads[72] of banana sales by the wholesalers in the modern system (using full cost approach)

Item	FCFA
Net Sales[73]	60000
Less cost of good sold[74]	45000
Gross margin[75]	15000
Less operating expenses:	
Transportation[76]	1720
Loading	-
Advertising and sales promotion[77]	500
Marketing administration[78]	2000
General administration[79]	500
Total operating expenses	4720
Net profit before taxes[80]	10,280

Source: Fonsah (1992) Field data.

Selling was done crudely. The retailers had no warehouses. Consequently, the truck-loads of reject bananas were dumped on the retailers' rented space, which served as warehouse, and were covered with banana leaves to prevent sunburn, rotting, and premature ripening.

---

[72] A truck-load is equivalent to three tractor-loads when overloaded. The 7 mt capacity trucks were usually preferred.

[73] Total revenue generated in the sale of a tractor-load of reject to the retailers was 20,000 FCFA. A seven metric tonne truck can carry three tractor-loads.

[74] The farm gate price of a tractor load of rejects was 15,000 FCFA or 5 FCFA/kg.

[75] Profit before operating costs.

[76] Estimated transportation cost to and from Douala=1000 FCFA. Township taxi to and from the market=220 FCFA. Town ship taxi from Likomba to the modern farm and back was 500 FCFA.

[77] Approximated cost incurred during contacts, promotion and negotiation with potential customers. Assume the wholesaler offered a drink, i.e., 250 by 2 = 500 FCFA.

[78] Hired labour who coordinated physical distribution at the farm level. This included round-trip transportation of 500 FCFA/day.

[79] Miscellaneous expenditure.

[80] Net returns before taxes even though taxes were never declared.

The retailer sold in stacks of 6-15 fingers for 25 FCFA depending on the season. Two seasons were identified, "La bonne saison" (the good season), which is the rainy season, and "La mauvaise saison" (the bad season), which is the dry season.

During the bad season, there were abundant supply of bananas and other food staples in the market, which facilitated substitution and eventually drove down prices as banana had to compete with other staples. The retailers sold about 10-15 fingers of banana for 25 FCFA, compared to 5-9 fingers during the favourable season when bananas and other agricultural food commodities were scarce. The number of fingers sold for 25 FCFA depended largely on the size and length of the bananas. The retailers claimed that profit was maximised only during the favourable season.

It was revealed that during the bad season the retailers were forced to maintain their "goodwill" by continuously buying from the wholesalers even though they claimed to be losing money. The reasons were twofold: (a) they maintain status quo in anticipation of the good season; and (b) if the status quo was disrupted, the wholesaler would, in retaliation, refuse to do business with those who breached their contracts during unfavourable season.

Empirically, it was found that the retailers preferred buying from the modern producer because: (a) the nearness of the modern firm (Tiko) to the market in Douala, (Appendix III), was approximately 70 km for a round trip, compared to 132 km from Ekona to Douala, and an average of 200 km from Penja to Douala; and (b) transportation cost from the farm to the market was cheaper in the modem than the semi-modem firms (Table 17.5).

Table 17.5 Cost of transporting a truck-load of banana from the farm to market

Capacity of Truck	Modern[81] (FCFA)	Semi-modern[82] (FCFA)
3 mt	25000	35000
5 mt	35000	45000
7 mt	45000	55000
10 mt	55000	65000

Source: Fonsah (1992) Field data.

[81] Prices for the modern system were negotiable due to the proximity to the Douala markets.

[82] Prices were firm due to the distance to Douala markets.

Table 17.6 Income and expense statement, FCFA/truck-loads[83] of banana sales by the wholesalers in the modern and semi-modern systems (using full cost approach)

Gross margin[86]	120 000	100 000
Less operating expenses:		
Transportation[87]	45 000	55 000
Loading[88]	3 750	-
Storage[89]	1 000	1 000
General administration[90]	500	500
Total operating expenses	50 250	56 500
Net profit before taxes[91]	69 750	43 500

Source: Fonsah E.G. (1992) Field data.

181

17.6 Market Structure

The wholesalers in the modern system had formed an oligopolistic market. Only three wholesalers, from different ethnic groups, were responsible for the purchasing and the distribution of all the rejected bananas (approximately 6,000 mts/yr) from the modern producer. The wholesale market was 100% dominated by women. The average educational qualification of the wholesalers was of the secondary school level. Entry into the wholesale cycle was almost impossible and political.

---

[83] A truck-load is equivalent to seven metric tonnes, but can carry upto 9 mt if overloaded. Three tractor-loads are equivalent to a truck-load.

[84] Total revenue generated in the sale of a tractor-load of reject bananas by the retailer to the final consumer.

[85] A tractor-load cost 20,000 FCFA and 10,000 FCFA in the modern and semi-modern systems, respectively.

[86] Profit before operating costs.

[87] Cost of hired truck from and to Douala (see Table 17.4).

[88] Cost of three hired loaders/truck in the semi-modern system. The producer absorbs this cost in the semi-modern system.

[89] Cost of rented space in the local market.

[90] This is spent for lunch as it takes six or more hours to load a truck.

[91] Net returns before taxes even though not usually declared

On the other hand, the banana retail market was highly segmented and was monopolised by the largest ethnic group in Cameroon, the Bamilekes from the Western Province. Consequently, penetration by non-ethnic members was extremely difficult. At the market level, there was also an oligopoly, whereby only a given number of retailers could sell reject bananas in a given segregated market. The main barriers to entry at this juncture were available space and ethnicity.

The existing retailers in each of the markets in this study had created an oligopolistic atmosphere by renting out all the spaces available in the agricultural food section of the local markets. Any forceful entry attempt was usually unsuccessful since the entrant had no place to display her reject bananas. The only possibility was to negotiate with one of the retailers, who might accept to either share her space or rent it out. But most of the tune, there was no room for negotiation especially if one belonged to another ethnic group.

Since farm land is almost non-existent in the Douala metropolitan area where the Bamilekes make up the majority, there has been a boom in reject banana trade for the following reasons:

(a) It is one of the staple foods for the Bamilekes;

(b) It is relatively cheaper than other food commodities;

(c) Banana is soft, thus easier and cheaper to cook in terms of time, gas and/or firewood consumption;

(d) It can be eaten either green (usually) or ripe (occasionally) and since most of the Bamileke households are thickly populated, no portion of the fruit is wasted;

(e) All the banana plantations are in the Douala agglomeration thus the supply is more or less consistent; *and*

(f) Irrespective of the social class (status), green banana is still cherished by this ethnic group.

Empirically, similar characteristics were identified in six out of the ten selected markets for the study. Surprisingly, in the Tiko, Mutengene, and Muea markets of the Fako Division as well as the Penja, Bouba III, and Mbanga markets of the Mungo Division, the market nichers were, at the same tune, the market leaders as illustrated in Fig. 17.11.

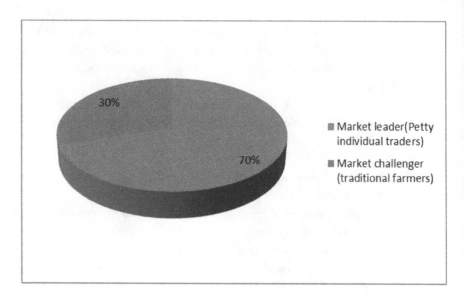

Fig 17.11

The Tiko market opens six days a week (Monday-Saturday), the Mutengene market twice weekly (on Wednesdays and Saturdays), while the Muea market opens twice weekly (Thursdays and Sundays). On the other hand, the Penja and Mbanga markets open every day, whilst Bouba III opens once a week (Thursdays only).

Approximately 70% of the bananas sold in these six markets were by individual petty traders (nichers) who were either relatives or Mends of workers from the semi-modern plantations (Mungo division), who took advantage of the newly implemented free banana policy by obtaining more than they could possibly consume, and then selling the excess. Some of the market nichers in the Fako division were selling stolen bananas from the modern plantation. None of the established retailers were identified in these markets. On the other hand, 30% of the products found were strictly in bunches whereas the nichers sold either in fingers, bands, or occasionally in bunches. Prices were determined by bidding until an equilibrium price was reached between the seller and the buyer.

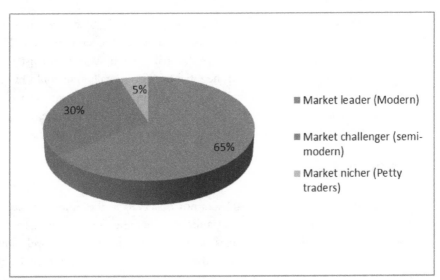

Fig 17.12

The four main markets in Douala metropolitan area, where approximately 95% of all the rejects were sold, were characterised by a different market structure (Fig. 17.12).

These markets were Deido, Bonaberi, Nkololo, and CiteSic. The markets operated on daily bases with a higher turnout and longer hours on non-working days such as Saturdays and Sundays. Douala has the two-shift working system, i.e. first shift starts from 7:30 to 12:00, with 12:00 to 14:30 being the break period; whilst the second shift is 14:30 to 18:00. As a result, consumers who could not attend the market in the morning, due to their work schedule, could do so during the lunch break, i.e. between 12 noon and 2:30 pm. The marketers, who open at 7:00 am, start closing for the day from 2:30 pm during regular working days, i.e. Monday-Friday. On Saturdays and Sundays, the situation was completely different. Consumers did their laundries, engaged in sports, attended church services, visited parents, and engaged in other activities which were impossible during working days. They attended the markets later in the afternoon. Consequently, market hours extended from 7:00 am to 6:00 pm on these days.

In these markets, 65% of the bananas sold came from the modern plantation, thus classifying the modern producer as the market leader. On the other hand, the market challengers were firms using the semi-modern production system that supplied 30% of the market share. The nichers in these markets were petty traders who bought in 50 kg size bags from the main retailers and resold in various isolated areas of the same market. They made up 5% of the market share. Traditional farmers were non-existent in these markets. Prices were seldom negotiable. Occasionally one could bid for an additional finger or two, usually referred to as "dash" or a gift, especially for regular clients. Sometimes the seller used the "dash" strategy on an identified new customer so as to render him/her regular in future.

17.7 Quality Considerations in the Domestic Market

Quality, in the strictest sense, was not observed in the domestic market. Quality was defined in terms of cleanliness, size, and finger length. In fact, consumers had mixed preferences as far as cleanliness, sizes, and finger lengths were concerned. Results of the consumer survey conducted La the ten major domestic markets during this study revealed that 87% of those who ate ripe bananas preferred small and clean (between 30-33 mm in diameter) sizes and 15-17 cm long (measured from the inner-whorl fingers). On the other hand, 72% of those who ate them green preferred sizes 34-38 mm in diameter and a length range of 18-22 cm. Furthermore, 63% of those who used bananas as animal feed also preferred the latter sizes and length but added that they would gladly settle for any available grade. Cleanliness, *per se,* mattered much to 90% of the consumers, especially those who ate them ripe, and to 45% of those who ate them green since the fingers were staked on the ground during sales.

It should be emphasised that fruits arrived at the domestic market, at the latest, twenty-four hours after harvesting, and were usually sold in less than a week. Consequently, not enough tune was allowed for major quality defects such as fusarium rot and crown moulds to appear. On the other hand, 84% of the local consumers cared little about such defects as red rust, over grade, maturity stain, chimera, etc. Even when present, they claimed that it did not affect the pulp or the taste. In cases of knife injuries, mutilated fingers, and split peels, the affected fingers or clusters were replaced by the seller - thus the problem was instantly solved.

17.8 Summary

An export trend analysis from 1978 to 1991 showed that export volume increased in the modern era, 1989-1991, compared to the semi-modern and traditional era, 1978-1988, respectively. Furthermore, 1991 was the best year in terms of total exportable output in more than a decade.

A monthly export and price trend of the four major producers of the preferential market: Cote d'Ivoire, Cameroon, Guadeloupe, and Martinique, showed that Cameroon received the least price for its product and was the least supplier in terms of volume in 1989.

A comparison of free-on-rail (FOR) price and export volume revealed that profit maximisation was impossible due to imperfect price timing. The Cameroon banana firms optimised export volume when world prices were unfavourable and vice versa. On the contrary, the price timing of their counterparts, Martinique and Guadeloupe, was perfect, whilst Cote d'Ivoire's export volume was constant irrespective of price trends.

An analysis of the domestic market showed three different distribution channels namely: the zero-level channel, whereby the producer sold directly to the consumer; the one-level channel, whereby the producer sold to the retailer, who in turn sold to the final consumer; and the two-level-channel, in which the merchandise left the producer to the wholesaler, retailer, and finally to the consumer. The zero and one-level-channels were common in both the traditional and semi-modern systems, whereas the two-level-channel was common in the modern system.

A marketing cost analysis revealed that the intermediaries (wholesalers and retailers) absorbed most of the costs incurred between the farm gate and the domestic markets.

A comparative income and expense statement using the full-cost approach showed that the intermediaries obtained the highest profit per truck-load of bananas in the modern system compared to the semi-modem system.

Bibliography

1. Abbott P. C. (1978). "Modeling International Grain Trade with Government controlled Markets." *AJAE*. 61, 22-31.

2. Acland, J. D. (1971). *East African Crops*. London: Longman, London.

3. Adams, M. R. (1978). "Small-Scale Vinegar Production from Bananas." *Trap. Science*, 20, 11-19.

4. Adamu, A.C. and E. G. Fonsah (1992). "Snail Control Measures in Cameroon Banana Plantation." *Del Monte Research Dept.* Douala, Cameroon.

5. Akobundu, I. O. (1987). *Weed Science in the Tropics*: Principles and Practices. New York: John Wiley and Sons.

6. Aldrich, R. J. (1984). *Weed-crop Ecology: Principles in Weed* Management. Breton Publishers, North Scituate, Massachusetts, 465pp.

7. Allen, R. N. (1978). "Spread of Bunchy-top Disease in Established Banana Plantations". *Australian Journal of Agricultural Research*, 29, 1223-33.

8. Almy, S. W., M. T. Besong and B. Bakia (1990). *Food Prices in South West Province: A Study of Food Crops in Twelve Markets*, Ekona; Institute of Agricultural Research.

9. Ambe, J. T. (1987). *Studies of Some Agronomic practices in a Cassava-based Cropping System in the South-West Province of Cameroon*. Unpublished Ph.D thesis, University of Ibadan, Nigeria.

10. Anon (1976). "Etude des Effects de la Temperature Sur lacroissance." *Fruits*, 31, 270-1.

11. Anthony, R. N. And A. G. Welsch (1981). *Fundamentals of Management Accounting* 3rd Ed. Richard D. Irwin, Hassewood, Illinois.

12. Arnold, D. R., M. L. Capella, and D. G. Smith (1983). *Strategic Retail Management.* Addison-Wesley Publishing Co. London.

13. Aubert, B. (1968). "Etude preliminaire des phenomenes de transpiration chez le bananier". *Fruits,* 23, 357-81.

14. Aubert, B. (1971). "Action du Climat sur le Comportement du Bananier en Zones Tropicales et Subtropicales." *Fruits,* 26, 175-88.

15. Badival, E. and K. G. Shanmugavelu (1978). "Effect of Increasing Rates of Potash on the Quality of Banana cv. 'Robusta'." *Potash Rev.* 24(8), 1-4.

16. Badran, A. M. (1969). "Controlled Atmosphere Storage of Green Bananas.[11] *US Patent* 3, 450, 542, 17 June.

17. Bale, M. D. And E. Lutz (1978). "Trade Restrictions and International Price Instability." *World Bank Staff Work. Paper No. 303. Oct.*

18. Bale, M. D. and E. Lutz (1979). "The Effect of Trade Intervention on International Price Instability." *AJAE,* 61, 512-16.

19. Bale, M. D. and E. Lutz (1981). "Price Distortions in Agriculture and their Effects: An International Comparison." *AJAE.* 63, 8-22.

20. Barker, W. G. (1959). "A System of Maximum Multiplication of the Banana Plant." *Trop. Agriculture,* 36, 275-84.

21. Basinger, A. J. (1931). "The European Brown Snail in California." *Bulletin 151, University of California Agricultural Extension Station, BULL.*

22. Beattie, B. R. and C. R. Taylor. (1985). *The Economics of* Production. New York: John Wiley & Sons.

23. Ben Meir, J. (1979). "A Case of Iron Excess in Canary Island Bananas". *Int. Banana Nutr. Newsl.* 1, 11-14.

24. Bequaert, J. C. (1950). Studies in the Achatininae, a Group of Giant Land Snails. *Bulletin of the Museum of Comparative Zoology,* 105(1): 1-216.

25. Berg, L. A, and M. Bustamante. (1974). "Heat Treatment and Meristem Culture for the Production of Virus-free Bananas." *Phytopathology,* 64, 320-5.

26. Besong M. and S. Almy (1990). "Farmers Land Preparation Methods in Southwest Province, Cameroon and Their Effect on Farm Labour Input." Paper presented at the *AFSRE 10th Annual Symposium,* Michigan State University, East Lansing, USA.

27. Besong, M. T. (1989). *Economics of Cassava Production in* Fako Division Cameroon: A Comparative Analysis of Local and Improved Cassava. Unpublished Ph.D Dissertation, Dept. of Agricultural Economics, Nsukka: University of Nigeria.

28. Bigman, D. and S. Rentlinger (1979a). "Food Prices and Supply Stabilisation: National Buffer Stocks and Trade Policies." *AJAE.* 61, 657-671.

29. Bigman, D. and S. Rentlinger (1979b). "National and International Policies Toward Food Security and Price Stabilisation." *Amer. Econ. Rev.* 69, 159-63.

30. Blake, C. D. (1972). "Nematode Disease of Banana Plantations." In: Webster, J. M. (ed), *Economic Nematology.* London: Academic Press.

31. Boehlje M. D. and V. R. Eidman (1984). *Farm Management,* New York: John Wiley and Sons.

32. Brady, Nyle C. (1974). *The Nature and Properties of Soils,* 8th Edition, New York: Macmillan Publishing Co., Inc.

33. Bredahl M. E., W. H. Myers, and K. J. Collins (1978). "The Elasticity of foreign Demand for US Agricultural Products: the Importance of Price Transmission Elasticity". *Am. J. Agric. Econ.,* pp58-63.

34. Bredahl, Maury E. (1976). "Effects of Currency Adjustments Given Free Trade, Trade Restrictions, and Cross Commodity Effects." *Dept. Agr. Econ. Staff. Paper.,* University of Minnesota, St. Paul.

35. Brignam, E. F. (1982). *Financial Management.* Theory and Practice, 3rd Ed. The Dryden Press, New York.

36. Buddenhagen, I. W. (1961). "Bacterial Wilt of Banana; History and Known Distribution." *Trap. Agriculture.* 38, 107-21.

37. Buddenhagen, I. W. (1960). "Strains of Pseudomonas Solanacearum in Indigenous Hosts in Banana Plantations of Costa Rica." *Phytopath,* 50, 660-4.

38. Bullock, R. C. And F. S. Roberts (1961). "Platynota Rostrana (walker): a Peel-Feeding Pest of Bananas." *Trap. Agriculture.* 38, 337-41

39. Burch, J. B. (1960). "Some Snails and Slugs of Quarantine Significance to die United States." *US Dept. Agr. Res. Ser.* 82(1): 1-70.

40. Burfisher, Mary E. (1984). *Cameroon - An Export Market* Profile. Washington DC: United States Dept. of Agriculture.

41. Butler, A. F. (1960). "Fertiliser Experiments with Gros Michel Banana." *Trap. Agric.* 37, 31-50.

42. Cann, H. J. (1964). "How Cold Weather Affects Banana Growing in N.S.W." *Agric. Gaz. N.S.W.* 75, 1012-1019.

43. Chalker, G. C. and D. W. Turner (1969). "Magnesium Deficiency in Bananas." *Agric. Gas. N.S.W.* 80, 474-476.

44. Chambers, R. G. and R. E. Just (1978). "A Critique of Exchange Rate Treatment in Agricultural Trade Models." *AJAE.* 61, 249-57.

45. Champion, J. (1976). "Quelques Problemes Production des Bananes Plantains." *Fruits,* 31, 666-8.

46. Champion, J., F. Dugain, R. Maignien, and Y. Dommergues (1958). "Les Sols de bananeraies et lew amelioration en Guinee, *Fruits,* 13, 415-462.

47. Charpentier, J. M., and P. Martin-Prevel (1965). "Culture Sur Milieu Artificiel Carences Attenees ou temporaires en el Majeurs, Carenre en Oligo-Element chez le Bananier." *Fruits,* 20, 521-557.

48. Charpentier, J. M. (1976). "La Culture Bananioaux iles Canaries." *Fruits,* 31, 569-85.

49. Chiang, A. C. (1984). *Fundamental Methods of Mathematical* Economics. 3rd Edition. New York: McGraw-Hill Book Company.

50. Chidebelu, S. A. N. D. (1980). *An Analysis of Factors* Affecting the Commercialisation of Limited Income Farms in Georgia. Unpublished Ph.D. thesis, University of Georgia, Athens, Georgia.

51. Chidebelu, S. A. N. D. (1980). Diffusion of Cooperative Mode of Production among Smallholder farmers of Nigeria." In O. Okereke (Ed.) *Cooperatives and The Nigerian Economy,* Nsukka: University of Nigeria.

52. Chidebelu, S. A. N. D. (1991). "Hired Labour on Smallholder Farms in South-eastern Nigeria." Issues in African Rural Development. Arlington, Virginia: *Winrock International Institute for Agricultural Development.*

53. Chidebelu, S. A. N. D. and G. C. Ezeronye, (1985). "Some Neglected Farm Management Problems affecting Food Farmers in Anambra State." *Journal of Rural Development and Cooperatives* 1(2): 131-194.

54. Christensen, R. C., R. K. Andrews, L. J. Bower, G. R. Hamermesh, and E. M. Porter (1982). *Business Policy.* Text and Cases. 5th Ed. Richard D. Invin, Inc., Homewood, Illinois 60430.

55. Chung, S. L. and S. P. Meyers (1979). "Bioproteins from Banana Wastes." *Development in Industrial Microbiology,* 20 723-32.

56. Clarke J. S. (1988). *Teach Yourself Chemistry,* Hodder and Stoughton.

57. Claude Culpin (1981). *Farm Machinery,* 10th Edition. London: The English Language Book Society and Granada.

58. Clemm, M. Bon (1964), "Agricultural Productivity and Sentiment on Kilimanjaro." *Econ.Bot.,* 18, 99-121.

59. CNCC (1983-1988). *Les Cahiers Statistiques.* Douala: Ministere de Transport (various issues).

60. Cochrane, Willard W. (1958). *Farm Prices: Myth and Reality.* University of Minnesota Press, pp 85-107.

61. Cochrane, Willard W., *A New Sheet of Music: How Kennedy's Farm Advisor Has Changed His Tune About Commodity Policy and Why.* University of Minnesota, pp 11-12.

62. Collins, K. J., H. W. Meyers, and M. E. Bredahl (1980). "Multiple Exchange Rate Changes and US Agricultural Commodity Prices." *AJAE.* 62, 656-665.

63. Conseil National Chargeurs du Cameroon (CNCC). *Etude sur le Transport des Bananes du Mongo.* Douala: Ministere de Transport, B21-J-31.

64. Cooke, G. W. (1982). *Fertilising for Maximum Yield,* 3rd Ed. London: ELBS/Collins.

65. Cronshaw, D. K. and J. L. Edmunds (1980). "Note on the Identification and Distribution of Moko disease in Grenada". *Trop. Agriculture,* 57, 171-2.

66. Croucher, H. H. and W. K. Mitchell (1940). "The Use of Fertilisers on Bananas." *J. Jamaican Agric. Soc* 44, 138-142.

67. Crowther, P. C. (1979), "The processing of Banana Products for Food Use." *Tropical Products Institute Publ.,* G122, London.

68. Cuille, J. (1950), Recherches sur le Charancon du Bananier, Cosmopolites Sordidus, Germ. Serie Tech. 4., *Inst. Fruit Agr. Col.,* 225pp.

69. Cunha, Antonio-Gabriel M. (Oct. 1984). "The Other Side of the Coin." *African Report,* 59-62.

70. Dale J. L. (1987). "Banana Bunchy-Top Virus: an Economically Important Tropical Plant Virus Disease. Advances in Virus Research". (Unpublished)

71. Dale, J. L. (1986). "Banana Bunchy-Top Virus: a Continuing Threat", In: *Banana and Plantain Breeding Strategies.* Inibap Aciar Proceedings No.21.

72. Dale, J. L., D. A. Phillips, and J. N. Parry (1986). Double Stranded RNA in Banana Plants with Bunchy-Top Disease. *Journal of General Virology.* 67, 371-75.

73. Darthenneq A. et al (1978). "Notes d'un Voyage d'etude dans Quelques Zones Bananieies d'Amerique Latine". *Fruits,* 33, 157-65.

74. Deacon, J. W. (1984). "Panama Disease of bananas in South Africa". *Horticultural Science 1,* 129-31.

75. Debertin, D. L. (1985). "Developing Realistic Agricultural Production Functions for Use in Undergraduate Classes." *SJAE,* December.

76. Debertin, D. L. (1986). Agricultural Production Economics New York: Macmillan Publishing Company.

77. Debertin, D. L. and Angelos Pagoulatos (1985)."Contemporary Production Theory, Duality, Elasticities of Substitution, the Translog Production Function and Agricultural Research." *Agricultural Economics Research Report* 40, University of Kentucky, Dept. of Ag-Economics, Lexington, Ky 40546-0215.

78. Debertin, D. L., Angelos Pagoulatos, and Garnett L. Bradford (1977). "Computer Graphics: An Educational Tool in Production Economics." *J. Ag-Economics*, 59, 573-576.

79. Dekle, G. W. (1969). "The Brown Garden Snail {Helix Aspersamuller)". Florida Dept. of Agric., Division of Plant Industry, *Entomology Circular* No. 83, Contribution No. 155.

80. Dekle, G. W. (1969). "The Brown Garden Snail *(Helix* Aspersamuller)". Entomology Circular No.83, April. Florida Dept. of Agri. Division of Plant Industry.

81. Deullin, R. (1970). "Refroidissement des Mains de bananes Condition dans des Caisses en Carton Ondul dans le Cas d'un Arrimage Compact". *Fruits,* 25, 583-91.

82. Devos, P. and G. F. Wilson (1983). "Associations du Plantain a d'autres Plantes Vivrieres II Autres Combinaisons avec le Mai's, le taro et le Manioc". *Fruits,* 38, 293-9.

83. Devos, P. and G. F. Wilson. (1979). "Intercropping of Plantain with Food Crops: Maise, Cassava, and Cocoyams". *Fruits,* 34, 169-74.

84. Dowdy, S. And S. Wearden (1983). *Statistics for Research.* John Wiley and Sons, New York.

85. Dupaigne, P. (1974). "A Propos de 1'Extraction d'un Jus de Banane, en Vue de la Production de la Biere de Banane." *Fruits,* 29, 821-2.

86. Dupaigne, P. (1967). "Le Controle de la Qualite des Bananes Sechees/Frwtt, 22, 27-9.

87. Eicher, C. K. and J. M. Staatz (1985). *Agricultural* Development in the Third World. Baltimore: The Johns Hopkins University Press.

88. Elliot Berg Associates (1983). *Agricultural Input Supply in* Cameroon, Vol. 1: A Report Prepared for the World Bank, Alexandria, Virginia.

89. Elz, D. (1987). "Agricultural Marketing Strategy and Pricing Policy." *A World Bank Symposium,* The World Bank, Washington DC.

90. Emory, W. C. (1980). *Business Research Methods.* Revised Edition. Homewood, Illinois, 60430.

91. Encyclopaedia Britannica, (1992), *Britannica World Data* Annual, Britannica Book of the Year: Chicago.

92. Epale, J. S. (1985). *Plantations and Development in Western* Cameroon, 1885-1975. New York: Vantage Press Inc.

93. F.A.O. (1972). The role of the Entrepreneur in Agricultural Marketing Development: ESMR: 72/Report, Rome: The German Foundation for Developing Countries.

94. F.A.O. (1982). *Committee on Commodity Problems.* Inter-Governmental Group on Bananas, 8th Session, CCP: BA 82/7.

95. F.A.O. (1983). *Commodity Review and Outlook,* Export Performance, 1966-1980. Rome: FAO Economic and Social Development Series, No.25.

96. F.A.O. (1983). *World Banana Economy: Statistical* Compendium, No.31. Rome: FAO Economic and Social Development Series.

97. F.A.O. (1986). *The World Banana Economy 1970-1984:* Structure, Performance, and Prospects. Rome: Economic and Social Development Series.

98. F.A.O. (1991). Committee on Commodity Problems. Inter-Governmental Group on Bananas, CCP:BA 91/6, Twelfth Session, Madeira, Portugal, 3-7 June.

99. Fanso, G. V. (1989). *Cameroon History for Secondary Schools* and Colleges, Vol. 1. London: Macmillan Publishers.

100. Feakin, S. D. (1971). Pest Control in Bananas (New Edition), London: Pans, 56 Gray's Inn Rd, London WCIX 8LU, England. Pans Maunal No. l.

101. Federal Reserve Bank of Kansas City (80-92): *Economic* Review, Various Issues.

102. Fergus, I. F. (1955). "A Note on Arsenic Toxicity in Some Queensland Soils." *Old. J. Agric. Set.* 12, 95-100.

103. Ffoulkes, D. et al (1978). "The Banana Plant as Cattle Feed: Composition and Biomass Production." *Trap. Anim. Prod.,* 3, 45-50.

104. Flinn, J.C. and J. M. Hoyoux (1976). "Le Bananier Plantain en Afrique. Estimation de son importance, rentabilit6 de sa recherche, Suggestions Economiques." *Fruits,* 31, 520-30.

105. Fourcroy and Vauquelin (1807). "Analyse du Sol de Bananier." Anns. Mus. d'Hist. Nat. 9, 301-302. In: Lahav and turner (1989). *Fertilising for High Yield Banana, IPI Bulletin 7, 2nd Edition.*

106. Foure, E. (1982), "Etude de la sensibilite Varietale des Bananiers et Plantains a Mycosphaerella Fijiensis au Gabon." (1), *Fruits,* 37(12), 749-770.

107. Foure, E. (1986). "Varietal Reactions of Bananas and Plantains to Black Leaf Streak Disease", in *Banana and Plantain Breeding Strategies,* Proceedings of an International workshop held at Cairns. Australia, Inibap Aciar No.21, pp110-113.

108. Foure, E. (1988b). Les Cercosporioses du Bananier et Leurs Traitements. Efficacites Comparees des differentes molecules fongicides sur Mycosphaerella fijiensis Morelet, agent de la maladie des raies noires des bananiers et plantains au Cameroun." *Fruits,* 43(1) 15-19.

109. Foure, E. (1988d). " Strategies de Lutte contre la Cercosporiose Noire des Bananiers et des Plantains Provoquees par Mycosphaerella fijiensis MORELET. L'avertissement biologique au Cameroun. Evaluation des possibilites d'amelioration." *Fruits,* 43(5), 269-274.

110. Foure, E. and A. M. Pefoura (1988a). "La Cercosporiose Noire de Bananiers et des Plantains au Cameroon *(Mycosphaerella fijiensis).* Contribution a l'etude des premieres phases de l'infection parasitaire. Mise au point de tests precoses d'inoculation sur plants issues de vitro-culture." *Fruits,* 43(2), 339-348.

111. Foure, E. et T. Lescot (1988c). "Variabilite Genetique des Mycosphaerella Infestes au Genre Musa. Mise en evidence de la presence au Cameroun sur bananiers et plantains d'une Cercosporiose *(Mycosphaerella musicola)* au comportement pathogene atypique." *Fruits.* 43(7-8), 407-415.

112. Foure, E. (1984). "Les Cercosporioses du Bananier et Leurs Traitements. Comportement des Varietes. Etude de la Sensibilite Varietale des Bananiers et Plantains a M. Fijiensis Morelet et de quelques Caracteristiques Biologiques de la Maladie des Raies Moires au Gabon (II)." *Fruits,* 39(6), 5-378.

113. Frankiel, L. (1989). "Les Achatinas aux Antilles". *Outre* Departmental de Documentation Pedagogigue, Circulaire.

114. Freiberg, S. R. and F. C. Stewart (1960). "Physiological Investigations on the Banana Plant: Factors which affect the nitrogen compounds in the Leaves." *Ann. Bot.* 24, 147-157.

115. G.I.E.B., (1990) *Le GIEB A Vmgt Arts,* (Exercise 1989), Paris, France.

116. Gamy, J. and J. P. Meyer (1975). "Recherche d'une Loi d'action de la Temperature sur la Croissance des Fruits du Bananier." *Fruits,* 30, 375-92.

117. Garcia, V., A. Diaz, E. Fernandez-Caldas, and J. Rabies (1976). "Factors que afectan a la asimilabilidad del potasio en los suelos de platanos de Tenerife." *Agrochimica* 12, 1-7.

118. Gassier, M. (1963). "The Banana in Geriatric and Low Calorie Diets." *Geriatrics,* 18, 782-6.

119. George, P. S. and G. A. King (1971). "Consumer Demand for Food Commodities in the United States with Projections for 1980". *Gianmini Foundation Monogr. 26* Davis: University of California, p.66.

120. Ghavami, M. (1976). "Banana Plant Response to Water Table Levels." *Transactions of the ASAE,* 19, 675-7.

121. Gibson, L. J., J. M. Ivancevich, and J. H. Donnelly Jr. (1982). *Organisation, Behaviour, Structure, Processes.* Irwin-Dorsey Ltd., Georgetown, Business Publications, Inc.

122. Oilman, L. J. (1982). *Principles of Managerial Finance.* 3rd Ed. Harper and Row Publishers, New York.

123. Gittinger, P. J. (1984). *Economic analysis of Agricultural* Projects. Baltimore: John Hopkins University Press.

124. Glasser, A. (1982). *Research and Development Management.* Prentice-Hall Inc. Englewood Cliffs, New Jersey 07632.

125. Gnanamanickam, S. S., T. S. Lokeswari, and K. R. Nandini (1979). "Bacterial Wilt of Banana in Southern India". *Pl. Dis. Reptr.,* 63, 525-8.

126. Godefroy, J., A. Lassoudiere, P. Lossois, and J. P. Penele (1978). "Actions du chaulage sur les Caracteristiques physio-chimiques et la Productivity d'un Sol tourbeux en culture bananiere." *Fruits* 33, 77-90.

127. Goewert, R. R. and J. J. Nicholas (1980). "Banana Peel Sugars as a Source of Foodstuff for Animals or Humans." *Nutrition Reports Int.,* 22, 207-12.

128. Goldstein M. and M. S. Khan (1976). "Large Versus Small Price Changes and the Demand for Imports." *IMF Staff Papers,* 23, 200-225.

129. Gowen, S. R. (1979). "Some Considerations of Problems Associated with the Nematode Pest of Bananas". *Nematropica* 9, 79-91.

130. Groupement D'Interet Economique Bananier, GIEB (1988). *Le* Marche National, Mondial et Communaute Economique Europeenne. Comptes Annuels Definitifs, Exercice Clos le 31 December, Paris, France.

131. Guignard, H. (1990). *L'Economie Bananiere.* Les Douze Marches de la Communaute Economique Europeenne et lews fournisseurs. Groupement D'interet Economique Bananier (G.I.E.B.), Paris, France.

132. Gunn, D. (1924). "The Brown and Grey Snails: Two Destructive Garden Pests." *Jour. Dept. Agr.* (Union of South Africa) Reprint No.42: 3-10.

133. Haarer, E. A. (1964). *Modern Banana Production.* London: Leonard Hill.

134. Hamilton, K. S. (1965), "Reproduction of Banana fromAdventitious Buds." *Trap. Agriculture,* 42, 69-73.

135. Harris, S., M. Salmon, and B. Smith (1978). *Analysis of* Commodity Markets for Policy Purposes, London: Trade Policy Research Centre.

136. Harrison, J. O. and C. S. Stephens (1966). "Notes on the Biology of *Ecpanteria Icasia,* a Pest of Bananas". *Am. Entomol. Soc. Am.,* 59, 671-4.

137. Harwood, R. R. (1979). Small Farm Development. Understanding and Improving Farming Systems in the Humid Tropics. Westview Press, Colorado.

201

138. Henderson, J. M. and R. E. Quandt (1980). *Micro-economic* Theory: A Mathematical Approach, 3rd edition. New York: McGraw-Hill Book Company.

139. Hodges, J. C. and E. M. Whitten (1977). Harbrace College Handbook. 8th Ed. Harcourt Brace Jovanovich, Inc. New York.

140. Holder, G. D., and F. S. Gumbs (1983b). "Effect of Water Logging on the Growth and Yield of Banana." *Trop. Agriculture,* 60, 111-16.

141. Hole J. (1984). "Price Stability and Public Policy". A Symposium Sponsored by the Federal Reserve Bank of Kansas City, Wyoming.

142. Hole, J. (1991). "Policy Implications of Trade and Currency Zones". *A Symposium Sponsored by the Federal Reserve Bank of Kansas City,* Wyoming.

143. Hord, H. H. V. and R. S. Flippin (1956). "Studies of banana weevils in Honduras". /. *Econ. Entomol.,* 49, 196-300.

144. Horngren, C. T. (1981). *Introduction to Managerial Accounting* 5th Ed. Prentice-Hall, New Jersey.

145. Houck, J. P. (1986). "The Basic Economics of and Export Bonus Scheme". *NCJAE.* Vol. 8, No. 2.

146. Hutson, J. C. and M. Park (1930). "Investigation of the Bunchy-top disease of Plantian in Ceylon". *Trop. Agriculturist,* 75, 127-40.

147. Hwang (1985). "Mass Screen for Fusarial Wilt Resistanceusing Banana Plantlets from Meristem Culture". *Fursarium Notes. 4,3.*

148. Hwang, S. C. and W. H. Ko (1986). "Some Clonal Variation of Banana and Screening for Resistance to Fusarium Wilt". *Banana and Plantain Breeding Strategies,* Inibap ACIAR proceedings, No 21.

149. Hwang, S. C., C. L. Chen, and F. L. Wu (1984). "An Investigation on Susceptibility of Banana Clones to Fusarial Wilt, Freckle and Marginal Scorch Disease in Taiwan". *Plant Protection Bulletin* (Taiwan). 26, 155-61.

150. Ingram, J. C. (1983). *International Economics*. New York; John Wiley and Sons, pp. 100-116.

151. Ingram, W. M. (1946). "The European Brown Snail *in* Oakland, California". *Bulletin of California Academy of Sciences,* 45(3): 152-159.

152. Inibap, (1986). *Banana and Plantain Breeding Strategies,* Brown Prior Anderson Pty. Ltd., Burwood: Victoria 3125, ACIAR No. 21.

153. Inibap, (1987a). *Genetic Improvement of Bananas and* Plantains and the Inibap Project on International MUSA Germplasm Exchange. Montpellier: France. 87/11, August.

154. Inibap, (1987b). *International Network for the Improvement of Banana and Plantain.* Annual Report, Montpellier: France. 88/03.

155. Inibap, (1988). *Nematodes and the Borer Weevil in Bananas,* Montpellier: France. 88/01, May.

156. Institut de Recherches sur les Fruits et Agrumes, IRFA (1981). Le Transport Interieur des Bananes - Rail ou Route?, Recherche D'une Amelioration des Conditions de mise en marche. CNCC. Douala: Republic Unie du Cameroun.

157. IRA Institute de la Recherche Agronomique (1989)."Programme de Recherche Bananiers et Plantains." *RAPPORTD'ACJWITES,* Nyombe, Cameroon.

158. IRAF (1978). *La Production et le Transport a Quai Bonaberi* de la Banane et de l'Ananas du Cameroun, Centre des Cultures Vivrieres et fruitiers, CNCC. Douala: Republic Unie du Cameroon.

159. Jackson, I. J. (1977). *Climate, Water, and Agriculture in the* Tropics, London: Longman.

160. Johnson, G. L. (1958). *Supply Function - Some Facts and* Notions in Acricultural Adjustment Problems in a Growing Economy (Ames: Iowa State University Press), p. 78.

161. Jones, U. S. (1982). *Fertilisers and Soil Fertility,* 2nd Ed., New Delhi: Prentice-Hall of India Private Limited.

162. Jose, P. C. (1981). "Reaction of different varieties of Banana against bunchy-top disease". *Agricultural Research Journal of Kerala* 19, 108-10.

163. Just, R. E., E. Lutz, A. Schmitz, and S. Turnovsky (1977). "The Distribution of Welfare Gains from International Price Stabilisation under Distortions". *AJAE. 59,* 652-61.

164. Kayisu, K. and L. F. Hood (1981). "Molecular Structure of Banana Starch," J. *Food Sci.,* 46, 1894-7.

165. Kayisu, K., L. F. Hood, and P. J. Vansoest (1981). Characterisation of Starch and Fibre of Banana Fruits." J. *Food Sci.,* 46, 1885-90.

166. Kenen, P. B. (1995). *International Economy.* New Jersey: Prentice Hall, Inc.
167. Kindleberger, Charles P. (1968). *International Economics.* Homewood, III.: Richard D. Irwin, Inc., pp.253-309, 569-587.

168. Knutson, Ronald D., J. B. Penn, and W. T. Boehm (1983). *Agricultural and Food Policy.* New Jersey: Prentice Hall Inc.

169. Kohls, R. L., and W. D. Downey (1972). *Marketing of Agricultural Products.* New York: MacMillan Publishing Co. Inc.

170. Kost, William E. (1976). "Effects of an Exchange Rate Change on Agricultural Trade." *AER,* 28, 99-106.

171. Koszler, V. (1959). "Banana for Infantile Diarrhoea." *Neue* Oesterr. Z. Kinderheilkd., 4, 212-14.

172. Kotler, Philip (1980). Marketing Management: Analysis, Planning, Implementation, and Control. 4th Edition. New Jersey: Prentice-Hall Inc.

173. Lacoeilhe, J. J., and J. Godefroy (1971). "Un Cas de Carence en Phosphore en Bananeraie." *Fruits 26,* 659-662.

174. Lacoeuilhe, J. J. And P. Martin-Prevel. (1971). Culture sunnilieu Artificial. 1. Carences en N, P, S Chez le Bananier: Analyse Foliaire. 2. Carence en K, Ca, Mg Chez le Bananier: Analyse Foliaire." *Fruits* 26, 161-167, 243-251.

175. Lahav, E. (1972). "Effect of Different Amounts of Potassium on the Growth of the Banana." *Trap. Agric.* 49, 321-335.

176. Lahav, E. (1973). "Phosphorus and Potassium Penetrability in the Soil and their Influence in a mature Banana Orchard." *Trap. Agric.* 50, 297-301.

177. Lahav, E. (1977). "L'Aprtitude de 1'Echantillonage du P6tiole a la determination des Teneurs en Mineraux de Bananier." *Fruits 32,* 294-307.

178. Lahav, E. and D. W. Turner (1989). *Fertilising for High field* Banana, 2nd Ed. Berne/Switzerland: Bulletin 7, International Potash Institute.

179. Lahav, E., M. Bareket, and D. Zamet. (1978). "The Effect of Organic Manure, KNO_3, and Combined Fertiliser on the Yield and Nutrient Content of Banana Suckers." *Proc. 8th Int. Collog. Plant anal, and Pert. Prob.,* Auckland, pp.247-254.

180. Lassoudiere, A. (1972). "Evolution de la Production Bananiere en Cote d'Ivoire." *Fruits,* 27, 829-53.

181. Lassoudiere, A. and P. Martin (1974). "Problemes de Drainage en Sols Organiques de Baneraire." (Agneby, Cote d'Ivoire). *Fruits,* 29, 255-66.

182. Leach, R. (1946). *Banana Leaf Spot (Mycosohaerella* Musicola) on the Gros Michel variety in Jamaica. Kingston, Jamaica: Government Printer.

183. Learner, E. E. and R. M. Stern (1970). *Quantitative* International Economics. Chicago, 111.: Aldine Publishing Company.

184. Lecoq, J. (1972). "L' Evolution de l'Economie Bananiere au Cameroun." *Fruits* 27, 677-96.

185. Lewis-Beck, M. S. (1980). *Applied Regression, An* Introduction: Series: Quantitative Applications in the Social Sciences. Sage Publications Paper 22, Newbury Park.

186. Liang, Li-feng (1989). "Banana Production and Research in Guangdong (China)", in: *Banana and Plantain R&D in Asia and The Pacific.* Inibap. Proceedings of a Regional Consultation on Banana and Plantain R&D Networking, Manila and Davao, 20-24 November.

187. Lii, C. Y., S. Chang, and Y. L. Young (1982). "Investigation of the Physical and Chemical Properties of Banana Starches". *J. Food Sci.,47,* 1493-7.

188. Locke, L. F.., W. W. Spiraduso, and J. S. Silverman (1993). Proposals that Work 3rd Ed. Sage Publication. International educational and Professional Publisher, Newbury Park.

189. Lowe, R. G. (1986). *Agricultural Revolution in Africa,* London: MacMillan Publisher.

190. Magee, C. J. (1927). "Investigation on the Bunchy-Top Disease of the Banana". Council for Scientific and Industrial Research. *Bulletin No. 30.*

191. Magee, C. J. (1936). Bunchy-Top Disease of Bananas. Rehabilitation of the Banana Industry in New South Wales. *Journal of the Australian Institute of Agricultural Science, 2,* 13-16.

192. Magee, C. J. (1948). "Transmission of Bunchy-Top to Banana Varieties." *Journal of the Australian Institute of Agricultural Science,* 14, 18-24.

193. Magee, C. J. And A. L. Fitzpatrick (1932). "Leaf fall of Bananas." *Agric. Gaz. N.S.W.* 43, 319-321.

194. Makouta, M. and M. M. Mbenoun (1983). *Etude Sur le* Transport Maritime de la Banana el des Fruits du Cameroun. Douala: CNCC, Ministere des Transports.

195. Maldonado, O., C. Ralz, and S.S. de Cabrera (1975). "Wine and Vinegar Production from Tropical Fruits". *J. Food Sci.,* 40, 262-5.

196. Mansfield, E. (1985). *Micro-economics.* Theory/Application. W. W. Norton and Co., New York.

197. Marchal, J. and P. Martin-Prevel (1971). "Les Oligos Cu, Fe, Mn, Zn dans de Bananier, Niveaus Foliaires et Bilans." *Fruits 26,* 483-500.

198. Marchal, J., P. Martin-Prevel, and Ph. Melin (1972). "Le soufre et le bananier." *Fruits* 27, 167-177.

199. Markman, T.R. H. and L. M. Waddell (1971). *10 Steps in Writing the Research Paper.* Barren's Educational series, Inc. New York.

200. Martin-Prevel, P. (1973). *Influence de la Nutrition Potassique sur les Fonctions Physiologiques et la Qualite de la Production chez Quelques plantes Tropicales.* 10th coll. Int. Potash Inst., pp233-248

201. Martin-Prevel, P. (1978). "Effects of Magnesium and Potassium Nutrition on Phosphorous Uptake and Redistribution in a Cultivated Plant, Musa sp. In: Plant Nutrition 1978." *Proc. 8th Int. Collog. Plant Anal. And Pert. Problems,* Auckland, pp329-338.

202. Martin-Prevel, P. and G. Montagut (1966). "Essais sol Plante sur Bananiers, 8. Dynamique de l'Azote dans la Croissance le et le Development du Vegetal. 9. Fonctions des Divers Organes dans l'Assimilation de P, K, Ca, Mg." *Fruits* 21, 283-294, 395-416.

203. Martin-Pr6vel, P. and J. M. Charpentier (1963). "Culture sur Milieu Artificiel: Symptomes des Carences en Six Elements Mineraux Chez le Bananier." *Fruits* 18, 221-247.

204. Massel, B. F. (1969). "Price Stabilisation and Welfare". Quart. J. Econ, 83, 284-98.

205. Matthews, G. A. (1979). *Pesticide Application Methods,* Essex, England: English Language Book Society (ELBS)/Longman.

206. Mbuagbaw, I.E., R. Brain, and R. Palmer (1987). *A History* of the Cameroon. Burnt Mill, England: Longman.

207. McCalla, A. F. (1966). "A Duopoly Model of World Wheat Pricing." *JFE.* 48, 711-27.

208. McCalla, A. F. (1967). "Pricing in the World Feed Grain Market." *Ag. Eco. Res.,* 19, 93-102.

209. Mead, A. R. (1961). *The Giant African Snail: A Problem in* Economic Malacology. Chicago, Illinios: University of Chicago Press.

210. Mead, A. R. And L. Palsy (1992). "Two Giant African Land Snail Species Spread to Martinique, French West Indies". *The Veliger.* 35(1): 74-77.

211. Melvin, Michael (1985). International Money and Finance, New York: Harper and Row Publishers, P. 146.

212. Menendez, T. and F. H. Loor (1979). "Recent Advances in Vegetative Propagation and their Implication to Banana Breeding." In: Proc. Fourth Conf. *ACORBAT,* edited by UPEB. Apartado 4273, Panama 5.

213. Meredith, D, S. (1970). "Banana, Leaf Spot Disease *(Sigatoka)* caused by *Mycosphaerella Musicola* Leech." *Phytopath.* Pap. 11, 147pp. Commonw. Mycol. Instit.

214. Meredith, D. S. and S. J. Lawrence (1970). "Black Leaf Streak Disease of banana {*Mycosphaerella Fijiensis*): Susceptibility of Cultivars." *Trop. Agriculture,* 47, 275-87.

215. Meredith, D.S., S. J. Lawrence, and J. D. Firman (1973). "Ascospore Release and Dispersal in Black Leaf Streak Disease of Bananas *(Mycosphaerella Fijiensis)." Trans. Brit. Mycol. Soc.,* 60, 547-54.

216. Mesling, J. H. L. (1974). "Longterm Changes in Potassium, Magnesium and Calcium Content of Banana Plants and Soils in the Windward Islands." *Trop. Agric.* Trin. 51, 154-160.

217. Michelini, Philip (1984). *Marketing in Cameroon,* (Overseas Business Report). Washington DC: US Department of Commerce.

218. Miles, M. B., and A. M. Huberman (1994). *Qualitative Data Analysis.* Sage Publicatsions, International Educational and Professional Publishers, Thousand Oaks.

219. Mills, R. L. (1977). *Statistics for Applied Economics and* Business. McGraw-Hill Book Company, New York.

220. Mirza, B. and R. Khalidy (1964). "Uptake of Magnesium through Foliar Spray in Banana." *W. Pakistan J. Agric. Res. 2,* 81-82.

221. Missingham, L. J. (1962). "Banana Yellows in North Queensland." *Queensland, Agric. J.* 88, 154-155.

222. Mitchell, H. S. (1968). *Coopers Nutrition in Health and* Disease. Philadelphia: J. B. Lippincott Co.

223. Murray, D. B. (1959). "Deficiency Symptoms of the Major Elements in the Banana." *Trop. Agric.* 36, 100-107.

224. Murray, D. B. (1960). The Effect of Deficiencies of the Major Nutrients on Growth and Leaf Analysis of the Banana." *Trop. Agric.* Trin, 37, 97-106, 119.

225. Murray, D. B. (1961). "Shade and Fertiliser Relations in the Banana." *Trop. Agric. Trin,* 38, 123-132.

226. Nayar, T. Gopalan (1962). *Banana in India.* The FACT Technical Society. Udyogamandal: Kerala State, India.

227. Ndubizu, T. O. C. and E. I. Okafor (1976). "Growth and Yield Pattern of Nigerian Plantains." *Fruits,* 31, 672-7.

228. Neba, A. S. (1987). *Modern Geography of the Republic of* Cameroon, 2nd Edition. Camden, N.J.: NEBA Publishers.

229. Nelson, N. D., M. Dobert, G. C. McDonald, J. McLanghlin, B. Marvin, and P. W. Moeller (1974). *Area Handbook for the United Republic of Cameroon.* Washington DC: US Government Printing Office.

230. Neuenschwander, P. (1988). "Prospects and Proposal for Biological control of *Cosmololites Sordidus* (Germar) *(cole optera Curculionidae)* in Africa". *Nematodes and the borer weevil in Bananas: Present status of research and outlook.* Proceedings of Inibap workshop, Bujumbura, Burundi, p.54-57.

231. Ngoh, J. V. (1987). *Cameroon 1884-1985: A Hundred Years of* History. Yaounde: Navi-Group Publication.

232. Nicholson, W. (1985). *Micro-economics Theory.* Chicago: The Dryden Press, Third Edition.

233. Orcutt, Guy H. (1950). "Measurement of Price Elasticities in International Trade." *Rev. Econ. and Stat.* 32, 117-32.

234. Orden, David (1980). "A Critique of Exchange Rate Treatment in Agricultural Trade Models: Comment." *AJAE.* 68, 990-97.

235. Oscatz, H. (1962). "Recent Findings and Experiences with the Fertiliser Treatment of the Banana." *Green Bull. No. 14,* Berlagsges, Ackerbau, Hannover.

236. Ostmark, H. E. (1974). "Economic Insect Pests of Bananas". Ann. Rev. Entomel, 15, 161-76.

237. Paris, K. (1979). The Lome Convention and the Common Agricultural Policy - Commonwealth Economic Papers, No.12. London: Commonwealth Secretariat.

238. Pegg, K. G. and P. W. Langdon (1986). *Banana and Plantain* Breeding Strategies. INIBAP Aciar Proceedings No. 2. pp 119.

239. Pindyck, R. S. And D. L. Rubinfeld (1981). *Econometric Models and Economic Forecasts.* 2nd Ed. McGraw-Hill Book Company, New York.

240. Pinochet, J. (1988a). "A Method for Screening Bananas and Plantains to Lesion Forming Nematodes". *Nematodes and the borer weevil in Bananas: present status of research and outlook.* Inibap Proceedings of a workshop held in Bujumbura, Burundi, 7-11 Dec., 1987, pp 62-65.

241. Pinochet, J. (1988b). "Nematodes Problems in Musa spp: Pathotypes of R. Similis and Breeding for Resistance". *Nematodes and the borer weevil in Bananas: present status of research and outlook.* Inibap Proceedings of a workshop held in Bujumbura, Burundi, 7-11 Dec., 1987, pp 66-70.

242. Pinochet, J. G. and R. H. Stover, (1980). "Fungi in Lesions caused by Burrowing Nematodes on Banana and their Root and Rhizome Rotting Potential". *Trap. Agriculture,* 57, 227-32.

243. Publication Manual of the American Psychological Association (1983). 3rd Ed. American Psychological Association, Washington DC, 20036.

244. Ramachandran Nair, P. K. (1979). "Intensive Multiple Cropping with Coconuts in India. Principles, Programmes, and Prospects". *Advances in Agronomy and Crop Science,* No.6, Verlay Paul Parey, Berlin and Hamburg, pp 147.

245. Reed, Michael R. (1980). "A Critique of Exchange Rate Treatment in Agricultural Trade Models: A Comment." *AJAE*. 62, 253-54.

246. Republique du Cameroun (1983). *Annuaire Statistiques*. Ministere du Plan et de L'Amenagement du Territoire. Yaounde: Direction de la Statistiques et de la comptabilite Nationale.

247. Reville, V.A. and L.A. Vegas (1967). La Enfermedad del "Moko" del Plantano en el Peru. Ministr. Agric. Serv. Invest. Promocion Agraria, *Boletin* Tecnico, N. 70, pp24.

248. Richert, G. H., W. G. Meyer, P. G. Marines, E. E. Harris (1974). *Retailing, Principles and Practices*. New York: McGraw-Hill Book Company.

249. Robinson, J. (1962). "The Influence of Interplanted Bananas on Arabica Coffee Yields." /. *Res. Rep. Coffee Res. Sta.*, Lyamungu, Tanganyika, 1961, 31-8.

250. Robinson, J. C. (1981). "F.I Water Requirements of Bananas. Farming in South Africa". *Banana F.I. Dept. of Agric. Fisheries*, Pictoria.

251. Roche, R. and S. Abreu (1983). "Control del Picudoi Negro del Platano *(Cosmopolites Sordidus)* Por la Hormiga Tetramorium Guinieensis". Ciencias Agric (Cuba), 17, 41-49.

252. Rojko, A. et. al. (1978). "Alternative Futures for World Food in 1985, Vol. I: World GOL Model Analytical Report," *Foreign Agric Econ. Rep* No. 146, Washington D.C. ESCS, USDA.

253. Ruttan, V. W. (1982). *Agricultural Research Policy*. University of Minnesota Press, Minneapolis.

254. Sanchez-Nieva, F., I. Hernandez and C. E. Bueso (1975). "Studies on the Freezing of Green Plantains (Musa Paradisiaca): 1, III, IV". *J. Agric.*, Univ. Puerto Rico, 59, 85-91, 107-14, 239-44.

255. Sanchez-Nieva, F.et al. (1970). "Effect of Zone and Climate on Yields, Quality and Ripening Characteristics of Montecristo Bananas Grown in Puerto Rico." J. *Agriculture,* University Puerto Rico, 54, 195-210.

256. Sankhayan, P. L. (1988). *Introduction to the Economics of* Agricultural Production, New Delhi: Prentice Hall of India Private Limited, New Delhi-11001.

257. Sanyal, A.K. et al. (1964). "Studies on Peptic Ulceration. I. role of banana in phenylbutazone induced ulcers". *Arch Int. Pharmacodyn Ther.,* 149, 393-400.

258. Sarah, J. L. (1990a). "Les Charancons des Bananiers". *Fruits,* Special bananes, pp 68-71.

259. Sarah, J. L. (1990b). "Les Nematodes et le Parasitisme des Raunes de Bananiers". *Fruits.* Special bananes 60-67.

260. Sassone, P.E. and W.A. Schaffer (1978). *Cost-Benefit Analysis:* A handbook Academic Press, New York.

261. Schuh, Edward E. (1974). "Exchange Rate and U S Agriculture. "*AJAE56,* 1-13.

262. Seelig, R. A. (1969). *Bananas: Fruits and Vegetable Facts and Pointers.* Washington, DC: United Fresh Fruit and Vegetable Assoc.

263. Sequeira, L. (1962). "Control of Bacterial Wilt of Bananas by Crop Rotation and Fallowing". *Trop Agriculture, Trin.,* 39, 211-17.

264. Shantha, H. and G. Siddappa (1970). "Accumulation of Starch in Banana Pseudostem and Fruit." J. *FoodSci.* 35, 74-7.

265. Shepherd, G. S. (1963). *Agricultural Price Analysis.* Ames Iowa: Iowa State. University Press.

266. Silberberg, Eugene (1978). *The Structure of Economics:* Mathematical Analysis, New York: McGraw-Hill Book Company.

267. Simmonds, N.W. (1966). *Bananas.* London: Longmans.

268. Simmonds, N.W.(1959). *Bananas: Tropical Agriculture Series,* London: Longmans.

269. Simon, H., K. Parris, C. Ritson and E. Tollens (1979). *Lome Convention and The Common Agricultural Policy Commonwealth Economic Papers:* N.12 Commonwealth Secretariat, London.

270. Spence, G. C. (1938). "Limicolaria as a pest". *Journal of Conchology.* 21:72.

271. Spinks, G.R. (1970). "Attitudes toward Agricultural Marketing in Asia, and the Far East." *Monthly Bulletin of Agricultural Economics and Statistics.* 19:(1).

272. Stanton, W. J. and R. H. Buskirk (1978). *Management of the Sales Force.* Fifth Edition. Homewood, Illinois: Richard D. Irwin, Inc.

273. Stover, H. (1962). "Fusarial Wilt (Panama disease) of Bananas and other Musa Species". *Phytopathology.* Pap. No.4. Commonwealth. Nucol. Institute.

274. Stover, R. H, E. S. Malo (1972), The Occurrence of Fusarial Wilt in Normally Resistant Dwarf Cavendish Banana". *PI. Dis. Kept,* 56, 1000-3.

275. Stover, R. H. (1972). "Banana, Plantain and Abaca Diseases."*Comm. Mycol. Inst.*

276. Stover, R. H. (1983). "The intensive Production of Horn-type Plantains (Musa AAB) with Coffee in Colombia. *"Fruits,* 38, 765-70.

277. Stover, R.H and N.W. Simmonds (1987). *Bananas: Tropical* Agriculture Series, 3rd Edition. New York: Longman Scientific & Technical.

278. Stover, R.H. (1980). "Sigatoka leaf spot of Bananas and Plantains." *Plant Disease, 64, 750.*

279. Subijanto (1990). "Country Paper Report on Banana and Plantain-Indonesia", in *Banana and Plantain R & D in Asia and the Pacific.* Inibap. Proceedings of a Regional Consultation on Banana and Plantain R & D Networking, Manila and Davao, 20-24, November 1989.

280. Subra, P. (1971). "Le 'Cableway¹, Mode de Transport des Regimes en Bananeraie." *Fruits.* 26, 807-17.

281. Swegle W. E. and P.C. Ligon (1989). *Aid, Trade, and Farm* Policies. Arlington, Virginia: Winrock International Institute for Agricultural Development.

282. Tarte, R., J. Pinochet, Gabrielli and J. Ventura (1981). "Differences in Population Increase, Host Preferences and Frequency of Morphological Variants among Isolates of the Banana Race for Radopholus Similis" *Nematropica.* 11, 43-52.

283. Teh-Wei Hu (1980). *Econometrics: An introductory Analysis.* University Park Press, London.

284. Tezenas du Montcel, H. du (1979). "Les Plantains du Cameroon. Propositions pour leur Classification et Denominations Vernaculaires." *Fruits,* 34, 83-97.

285. Tezenas du Montcel, H. du (1987). *Plantain Bananas:* The Tropical Agriculturalist. London: MacMillan Publishers.

286. The World Bank (1980). *Poverty and Human Development.* Oxford University Press.

287. The World Bank (1982). *Accelerated Development in* Sub'Saharan Africa, Washington DC: World Bank, April.

288. The World Bank (1983). *World Development Report,* Washington, DC.: The World Bank.

289. The World Bank (1991). *Agricultural Technology in Sub-Saharan Africa.* A Workshop on Research Issues No. 126, Washington, DC.

290. The World Bank Atlas (1992). *25th Anniversary Edition.* The World Bank, Washington, DC.

291. Tibaijuka, A. K. (1985). "Women in the Banana Industry in Tanzania" In: International Cooperation for Effective Plantain and Banana Research, Proceedings of the Meeting, Abidjan -Cote d'Ivoire, 27-31 May, pp. 193-195.

292. Ticho, R.J. (1970). "The Banana Industry in Israel". Jerusalem, Israel: Ministry of Agriculture.

293. Tomek W.G. and K. L. Robinson. (1981). *Agricultural Product Prices.* New York: Cornell University Press, Ithaca.

294. Turner, D. W. (1968). "Micro Propagation of Bananas." Agric. Gaz. N.S.W. 79, .235-6.

295. Turner, D.W., and E. Lahav. (1983). "The Growth of Banana Plants in Relation to Temperature. *"Aust J. Plant Physiol,* 10, 43-53.

296. Turner, D.W. and in J.H. Bull. (1970). "Some Fertiliser Problems with Bananas.*"Agric, Gaz, N.S.W. 81,* 365-367.

297. Turner, D.W., and B. Barkus. (1980). "Plant Growth and Dry Matter Production of the 'Williams' Banana in Relation to Supply of Potassium, Magnesium and Manganese in Sand Culture." *Scientia Hortic.* 12, 27-45.

298. Tumovsky, S. J. (1977). "The Distribution of Welfare Gains from Price Stabilisation of some Theoretical Issues. *"Paper presented at the Ford Foundation Conference on Stabilising World Commodity Markets,* Airlie, VA.

299. Twyford, I. T. and D. Walmsley (1973). "The Mineral Composition of the Robusta Banana Plant, Methods and Plant Growth Studies". *Plant and Soil,* 39, 227-243.

301. Twyford, I. T. and D. Walmsley (1974a). "The Concentration of Mineral Constituents: Uptake and Distribution of Mineral Constituents". *Plant and Soil.* 41, 459-491.

302. Twyford, I. T. and D. Walmsley (1974b). "The Application of Fertilisers for High Yields. *Plant and Soil;* 41, 493-508.

303. Twyford, I.T. (1967), "Banana Nutrition: a Review of Principles and Practice." *J. Sci. Fd. Agric 18,* 177-183.

304. United Republic of Cameroon (1976). *Fourth Fifth Year Economic, Social and cultural Development Plan of Cameroon,* 1976-81. Yaounde: Ministry of Economy and Planning.

305. United Republic of Cameroon (1980), Bilan Diagnostique du Secteur Agricole de 1960 a 1980.Yaounde: Ministry of Agriculture.

306. United Republic of Cameroon (1981a). L'Essential sur le Ve Plan Quinquennal de Developpement Economique, Social et Culture!, Yaounde^. Ministry of Economy and Planning.

307. United Republic of Cameroon (1981b). *Draft Fifth Five-Year Economic, Social and Cultural Development Plan of the United Republic of Cameroon.* Yaounde: Ministry of Economy and Planning.

308. United Republic of Cameroon (1984). *Cameroon, Economy,* Demographics. Washington DC: Embassy of Cameroon.

309. US Agency for International Development (1983). "The Tortoise Walk: Public Policy and Private Activity in the Economic Development of Cameroon". *AID Evaluation Special Study, No. 10.*

310. US Department of Agriculture (1982). Sub-Saharan Africa: *Review of Agriculture in 1982 and Outlook for 1983. Supplement 7 to WAS-31.* Washington DC: Economic Research Service.

311. US Department of Agriculture (1985). *World Indices of Agricultural Food Production. Washington,* DC: USDA.

312. US Department of Agriculture (July, 1985). *Sub-Saharan* Africa (Outlook and Situation Report), Washington DC: USDA.

313. US Department of Agriculture, USDA (1963). "Composition of Foods". *Agric. Handbook 8,* US Gov't. Print. Off., Washington, DC.

314. Vadivel, E. And K. G. Sharimugavelu (1978). "Effect of Increasing Rate of Potash on the Quality of Banana cv. Robusta! *Potash Rev.* 24(8), 1-4.

315. Varian, Hal R. (1978). *Microeconomic Analysis,* New York: W. W. Norton and Company.

316. Viquez, F., C. Lastreto, and R. D. Cooke (1981). "A Study of the Protection of Clarified Banana Juice using Pectinolytic Enzymes." *J. Food Technology,* 16, 115-25.

317. Walmsley, D. and I. T. Twyford (1968). "The Uptake of P by the 'Robusta' Banana." *Trop. Agric.* Trin. 45, 223-228.

318. Walmsley, D. and I. T. Twyford (1976). "The Mineral Composition of the Banana Plant. Sulphur, iron, manganese, baron, zinc, copper, soduim and aluminium." *Plant Soil* 45, 595-611.

319. Walmsley, D., I. Twyford, and I. S. Cornforth (1971). "An Evaluation of Soil, Analysis Methods for Nitrogen Phosphorus, and Potassium Using Banana." *Trop. Agric.* Trin. 48, 141-155.

320. Wannacott, R. J. And T. H. Wannacott (1979). *Econometrics.* 2nd Ed. John Wiley and Sons, New York.

321. Wardlaw, C. W. (1972). *Banana Diseases: including Plantains* and Abaca, 2nd edition, London: Longman.

322. Warner, R. M. and R. L. Fox. (1977). "Nitrogen and Potassium Nutrition of the Giant Cavendish banana in Hawaii." *J. Amer. Soc. Hon. Sci.*, 102, 739-743.

323. Waugh, F. V. (1944). "Does the Consumer Benefit from Price Instability?" *Quart. J. Econ.* 58, 603-14.

324. Wickham, T. (1985). *Irrigation Management.* Research from Southeast Asia, Agricultural Development Council Inc., Kampangsaen, Thailand.

325. Wilson R. J. (1975). The International Market for Banana Products for Food Use." *Tropical Products Institute Publ,* G103, London.

326. Wilson, G. F. (1976). "Le Plantain Dans les Systemes de Culture des Tropiques Humides". *Fruits,* 31, 517-519.

327. Wilson, G. F. and W. E. Takacs (1976). "Differential Response lags to Price and Exchange Rate Change in the Foreign Trade of Selected Industrial Countries." *Paper presented at the Annual meeting of Allied Social Sciences Association, Atlantic City,* NJ. 16-18.

328. Wilson, G. F. (1983). "Production de Plantains: Perspectives pour Ameliorer la Situation Alimentaire sous les Tropiques." *Fruits,* 38. 229-239.

329. Wrigley, G. (1981). *Tropical Agriculture The Development of Production,* 4th Ed., Burnt Mill, England: English Language Book Society/Longman.

330. Zimdahl, R. L. (1980). *Weed-Crop Competition:* A Review, International Plant Protection Centre, Oregon State University, Corvallis, Oregon, 197pp.

331. Zwart, A. C. and K. D. Meilke (1979). "The Influences of Domestic Pricing Policies and Buffer Stocks on Price Stability in the World Wheat Industry." *AJAE*. 61, 434-47

Appendix I

German Plantations in West Cameroon, 1885-1914.

Company	Founder	Share capital	Year	Has	Branches	Locations
West Africanische pflanzungs Gesellschaft	Jantzen und Thormahlen	£105,000	1885		Jantzenhof Thormahlenfield Dollmanshohe Extensions: Wernerfelde Retzlafelde Scharlachfelde Merckefelde Wete-wete	Bibundi, approx. 50 kms from Limbe Mukondange, approx. 10 kms from Limbe

Company	Owner/Manager	Capital	Year	Notes	Location
Kamerun Land-und Plantagen-Gesells-chaft	Woemann Commanditge-Sellschaft assisted in the creation. Actual ownership to Woermann on 6th July, 1906 and May 1909, respectively		1885		Man-O'War bay (Kriegschiff Hafen)-Mabetta, River Ombe, Tiko Plain and Mungo River
WestFri-kanische Pflansungs Gesellschaft Victoria (WAPV)	Gunther Dr. Max Esser Manager	£105,000	1897	Victoria branch consists of eleven WAPV farms	Victoria, Krater, Mitel, Ngeme, Limbe, Buana, Bosumbu, Ebongo, Wotutu, Sachsenhof and Mokunda Lysoka, Lysoka. Molyko, Tole, Moli Bulu and Bolifamba.
Prinz Alfed Pflansung (WAVP	Gunther		1908	Missellele: Malende Farm	Missellele: Tiko-Douala Rd Malende: Kumba Rd

Idenau Pflanzung Gessellschaft	Ferdinand Scipio and William Scipio	£50,000	1898 1912		The branch had five farms: Linkfluss, Rechfluss, Scipio, Clotilde, Soden.	North Bibundi
Debundscha Pflanzung	Paul Geuger Gunnar Linnel Hullmuth Von Olttzen Georg Waldau	£11,000	1902			Debundscha (wettest areas in Africa)
Plantage Oechelihausen	Dr. W. Von Oechel hausen and Prof. Dr. A. Von Oechel hausen		1898 but developed in 1904			4 kms of Victoria Bibundi Road
Molive Pflanzung Gesellschaft	Sholto Douglas	£100,000	1899	12826	Dibongo	Covered Ombe, N'sonne Moliwe plantation and the present Mondoni palm est Near Edea W. Cameroon
The Kantschuk Pflanzung "Meanja"		£45,000	1903	5570		Meanja River (Kumbia Road)

Company	Owner	Price	Year	Number	Composition	Location
Kamerun Kantschuk Compagnie Aktiengesellschaft (KKC Mukonje)			1909		Composed of six farms: Mungo, Laduma, Egua, Kompenda, Bekili, Gomalinge	Mukonje (present meme division)
Deutsche Kantschuk Atiengesellschaft (Ekona)	Esser Oechel hausen	£125,000	1907-1913		Divided into 8 farms: Lysoka, old koke, Musaka Mamu, Povo, small Mpondo big Mpondo	Kumba Road
Hilferts Pflanzung or Mungonge Estate	Herr Otto Hilfert		1907	2500		15 kms north of Idenau
Afrikanische Frucht Compagnie (AFC)[92]			1907-1912	4942	Local banana variety	West of Tiko Plain
Tiko Pflanzung M.B.H.	Otto Holtforth[93] Herr J. Rohricht Herr F. Stein hausen		2e+07	2e+08	Holtforth Estate Likumba Tiko/Likumba Longstreet Njoke	Tiko Tiko Tiko Victoria Kumba Rd

Estate	Surveyors/Persons	Year	Area	Notes	Location
Mundame Estate (Bremer Gesellschaft Nord-West Kamerun)	Bremer		0.5	Acquired two more pieces of land in Kumba village	Kumba
Essosong Estate (Tabacco land) (Bremer Tabakbau Gesellschaft) "Bokossi" M.B.H.	Bremer	1910		Subleased Kamerun Eisenbahn Gesellschaft (KGE)	Essosong Meme Division
Ombe River Estate	Surveyors Herr Wessel and Herr Christoph Rein	1912	242		Ombe
The Deutsch Westafrikanische Handells Gesellschaft			14000	Bavo-Mbonge Estate further acquired Ndian, Illoani, Rio del Ray	Ndian Division
Gesellschaft Sud kamerun (GSK)	Dr. Scharlach and Sholto Douglas	1898	from 7m down to 1.5m has		South and Southeast part of Southwest Province

Gesellschaft Nord West Kamerun (GNWK) Gesellschaft	Dr. Scharlach and Sholto Douglas	4-10 million Marks	1899	4.5m has	West and North-western part of Southwest Province

Source: Epale, J.S. *Plantations and Development in Western Cameroon*, 1885-1975. Vintage Press, Inc., New York, 1985 and tabulated by authors.

[92] The company first started banana production in 1907 using the local variety. Large-scale production and marketing began in 1912 after the introduction of 'Gros Michel' variety in 1910. This company also built the Tiko wharf in Keka Island in 1913.

[93] Otto Holtforth later acquired Rohricht and Steinhausen farms, respectively, and all became the Tiko Pfanzang (MBH).

Appendix II

Advantages of irrigation systems used in banana plantations

Overhead (above canopy) sprinkler	Undertree (below canopy) sprinker	Ground drip
1. Comparatively infrequent failures and durability.	1. High pressure, low volume energy consumption compared to overhead and an improved water dispersal.	1. Most efficient in energy consumption because of low pressure and low volume.
2. The development of micro-organisms that affect plant/banana leaves are minimised due to reduced wetting time compared to undertree system.	2. Reduced short-term operation cost.	2. Most efficient water management.
	3. Water utilisation and permeability rates are better managed in average to fine textured soils.	3. No development of micro-organisms caused by wetness.

| | 4. Daily plant water requirement can easily be regulated. |
| | 5. Most efficient input application based on planting pattern. |

Source: Stover R.H. and Simmond N.W. (1987) Bananas. Tropical Agricultural Series, 3rd Edition. New York: Longman, Scientific and Technical.

Disadvantages of irrigation systems used in banana plantations.

Overhead (above canopy) sprinkler	Undertree (below canopy) sprinker	Ground drip
1. It is high pressure, high volume, thus consumes the most energy.	1. Chemical treatment may be washed off depending on plant size and variety, thus exposing the leaves to various bacteria/fungi attacks.	1. Drippers are subject to clogging and require expensive filtration systems and periodic chemical treatment of water.
2. If uncontrolled, it washes off chemicals on treated leaves.	2. Uneven distribution of water especially in highly populated areas.	2. The drippers must be flushed frequently to prevent clogging especially when Urea is used.

| 3. Inconsistent water distribution. | 3. Vulnerability to thieves by children (sprinklers) and destruction of Pes and PVCs by replanters, pruners, and other field workers. |
| 4. Excessive water loss especially on heavy soils with inferior permeability rate and from evaporation. | 4. Installation in a conventional planting pattern or may be rather difficult. |

Source: Stover R.H. and Simmond N.W. (1987) Bananas. Tropical Agricultural Series, 3rd Edition. New York: Longman, Scientific and Technical.

Appendix III

Map of Cameroon showing Provincial Boundaries, Headquarters, and Export Banana Production Areas

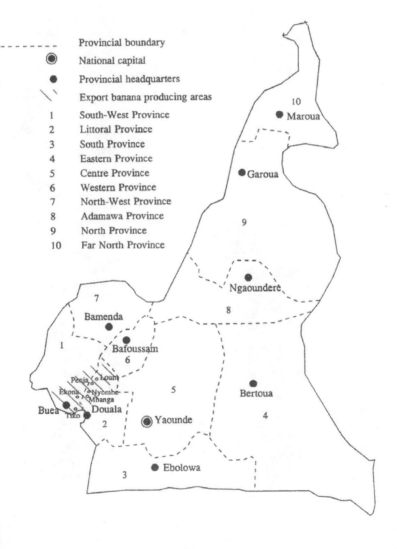

Appeandix IV

Cameroon in the African Continent

Appendix V

Summary of Deficiency Symptoms

Age of leaf	Symptoms on blades	Additional Symptoms	Element
All ages	Uniform paleness	Pink petioles	N
		Midrib curving (weeping, drooping)	Cu
Young leaves only	Whole leaf yellow-white		Fe
		Thickening of secondary veins	S
	Streaks across veins	Leaves deformed (blades incomplete)	B
	Stripes along veins	Reddish colour on lower side of youngest leaves	Zn
	Marginal chlorosis	Thickening veins, necrosis from margins inward	Ca
Old Leaves only	Sawtooth marginal chlorosis	Petiole breaking. Bluish-bronze colour of young leaves.	P
	Chlorosis in midblade, leaves madrib and margins remain green	Chlorosis limit not clear, pseudostem disintegrating	Mg
	Blade dirty yellow green		Mn
	Yellow-orange chlorosis	Leaf bending, quick leaf desiccation	K

Source: Lahav, E. and D.W. Turner (1989), *Fertilising for High Yield Banana,* 2nd Ed. Berne/Switzerland: International Potash Institute.

Appendix VI

Glossary of Terms

Aerial spray The use of either an aircraft or helicopter with microair to spray fungicides on banana plants/leaves.

Aerial tractor A machine that moves on the cableway and used to haul banana bunches from the field to the packing station. It is capable of hauling more than 100 bunches for very long distances. The operator sits on it without touching the ground.

Bagging The covering of the bunch with a polythylene bag to protect it from birds, insects, dirt, and for temperature control.

Banavac A technique of packing banana clusters in a thin polyethylene bag inside the box, usually for long voyages, to minimise the spread of disease on transit.

BCUF The Bakwerie Cooperative Union of Farmers Ltd.

Blowdown Fallen plant caused by wind. It is either uprooted, snapped off or broken midway.

Borer An insect whose larvae bores holes in the rhizome, killing young plants and increasing the tipovers of older plants.

Box-stem-ratio The number of boxes that can be packed from a stem or a bunch.

Bract The burgundy-colored part of the inflorescence that protects the emerging hands. They all fall off when the fruit is ready for bagging.

Bud The male part of the inflorescence.

Bunchy Top Virus A disease that affects the plant.

Calibration Measuring the diameter of the fruit.

CDC Cameroon Development Corporation.

CIF Cost, Insurance, Freight.

Cluster A hand of banana or plantain divided into two or more smaller parts.

COOPLABAM Cooperative des Planteurs des Bananes.

Conn The underground part of the banana or plantain plant that develops roots and serves as food storage and a reproductive organ.

Cultivar Different varieties of banana of the same biological origin.

Debudding The cutting of the male bud during bagging.

Deflowering The removal of the flowers, either in the field to protect the bunch from Cigar-end infection, or in the fruit patio before dehanding.

Dehanding The removal of individual hands from the bunch leaving only the stalk.

Deleafing The cutting off of broken, disease affected leaves, and either cutting or trimming leaves touching the bunch.

DMC Del Monte Cameroon S.A.R.L.

Double row A planting pattern with two lines.

Doubles The technique of leaving two daughters for production units after the mother plant has been harvested. It also refers to a plant that breaks in the midway after a windstorm.

Drains A gutter that facilitates the removal of excess water from the soil in the plantation.

Earthing Up The covering of the conn part of the plant with soil when the roots are exposed.

Ecpanteria sp The family of a peel eating caterpillar.

Ethylene A gas (product) used in ripening bananas.

Fallow A technique of improving soil fertility by abandoning a piece of land uncultivated for a year or more.

FAO Food and Agricultural Organisation of United Nations.

FED European Development Fund.

Finger A single individual banana fruit.

Fusarial wilt A disease that affects die plant. It is locally called Panama disease.

Herbicide Chemicals used for weed control.

In vitro The multiplication of seeds in die laboratory.

Inflorescence The burgundy colored pan of the unfolded bunch.

KMBAP International Network for the Improvement of Banana and Plantain.

IRA Institute of Agronomic Research.

K/F of CPMS Ltd Kumba Federation of Cooperative and Produce Marketing Societies.

Lome Convention A convention that provides preferential tariff treatment to banana producing countries in Africa, the Caribbean and French oversees exporting to the EEC.

Mat A complete plant.

Meristems Same as *in vitro*. The multiplication of seeds in die laboratory

Musa acuminate Origin of the edible banana.

Musa Balbisiana same as Musa acuminata.

OCB Organisation Camerounaise de la Banane.

Peduncle The distance between the crown and the beginning of the pulp. It is sometimes referred to the stalk.

Peeper The initial view of the emerging inflorescence in an upright position.

PFPC The Progressive Farmer's Produce Entreprise.

Phenology The study of plant behaviour i.e. the time required from planting to shooting, the number of leaves produced, etc.

PHP Plantation de Haute Penja.

Plant crop New plant that has not either attained harvesting stage, or has been harvested only once.

Pomology A study of the behaviour of the fingers, i.e. the length, the size and the curvature, to determine quality specifications.

Propping The protection of the pseudostem from wind damage by either tying with a twine or supporting with bamboos.

Pruning The elimination of unwanted suckers, leaving the most vigorous ones based on the location and position of adjacent plant, as a production unit.

Pseudostem The huge elongated part of the banana plant.

Replanting The replacement of a missing plant in an established plantation.

Ratoon The second and subsequent harvest of a banana plant.

Rhizome same as conn.

SAP Societe Agricoles de Prevoyance.

SDIBC Syndicat de Defense des Int6ret Bananiers du Cameroun. This was the first Cameroon banana producer's union.

Seedbeds A nursery or segmented part of a plantation used to multiply seeds.

Sigatoka A disease that affects the plant, mostly the leaves.

Snap off When the rhizome breaks as a result of windstorm causing the plant to fall.

SPNP Societ6 de Plantation de Nyombe-Penja.

Sucker The daughter or follower of a mother plant.

Sword sucker The daughter of the mother plant with small leaves usually used as planting material or maintained for production unit.

UBM Union Bananiere du Mungo.

UFC United Fruit Company.

UGECOBAM Union Gendrale des Cooperative Bananiers du Cameroun.

Uprooting When the whole plant topples including the root system during a wind storm.

USDA United States Department of Agriculture.

Water sucker The daughter of a harvested/destroyed mother plant, usually having broad leaves.